# Silane Coupling Agents

SECOND EDITION

# Silane Coupling Agents

## SECOND EDITION

## Edwin P. Plueddemann

*Scientist Emeritus*
*Dow Corning Corporation*
*Midland, Michigan*

PLENUM PRESS • NEW YORK AND LONDON

Library of Congress Cataloging-in-Publication Data

Plueddemann, Edwin P.
    Silane coupling agents / Edwin P. Plueddemann. -- 2nd ed.
        p.    cm.
    Includes bibliographical references and index.
    ISBN 0-306-43473-3
    1. Silane compounds.  2. Plastics.   I. Title.
QD412.S6P5  1991
668.4'16--dc20                                              90-21011
                                                                CIP

© 1991, 1982 Plenum Press, New York
A Division of Plenum Publishing Corporation
233 Spring Street, New York, N.Y. 10013

Printed in the United States of America

# Preface to the Second Edition

Much progress has been made in the last 8 years in understanding the theory and practice of silane coupling agents. A major advance in this direction was the measurement of true equilibrium constants for the hydrolysis and formation of siloxane bonds. Equilibrium constants for bond retention are so favorable that a silane coupling agent on silica has a thousandfold advantage for bond retention in the presence of water over an alkoxysilane bond formed from hydroxy-functional polymers and silica. In practice, the bonds of certain epoxies to silane-primed glass resist debonding by water about a thousand times as long as the epoxy bond to unprimed glass.

Oxane bonds of silane coupling agents to metal oxides seem to follow the same mechanism of equilibrium hydrolysis and rebonding, although equilibrium constants have not been measured for individual metal–oxygen–silicon bonds. This suggests, however, that methods of improving bond retention to glass will also improve the water resistance of bonds to metals. Modification of standard coupling agents with a hydrophobic silane or one with extra siloxane cross-linking have improved the water resistance of bonds to glass and metals another hundredfold over that obtained with single coupling agents.

Bonds of polymers to silane-treated surfaces may involve simple reaction of the polymer with functional groups of the silane, but most often involve an interdiffusion of siloxane oligomers into the polymer with subsequent cross-linking to form interpenetrating polymer networks in the interphase region. This mechanism correlates with the observation that a silane primer that bonds two different polymers to glass can also be used to bond the two polymers to each other.

A simple screening test has been used for evaluating silane primers in bonding many thermosetting and thermoplastic polymers to glass and metals. Ratings from these tests provide a guide for formulators in developing optimum bonding primers.

Bonding through silane coupling has been disappointing in processes, such as extrusion or injection molding, that involve high shear in the melt. Bonds between coupling agent and polymer are made and broken during the time of high-temperature shear, so that the full value of coupling agents are not retained. A new method has been proposed that bonds silane-treated particulate fillers to polymers through ionomer linkages. Under conditions of high-temperature shear, the ionomers are fluid, but they set to bonded structures when cooled to room temperature.

Edwin P. Plueddemann

Preface to the First Edition

It has been rumored that a bumblebee has such aerodynamic deficiencies that it should be incapable of flight. Fiberglass-reinforced polymer composites, similarly, have two (apparently) insurmountable obstacles to performance: (1) Water can hydrolyze any conceivable bond between organic and inorganic phases, and (2) stresses across the interface during temperature cycling (resulting from a mismatch in thermal expansion coefficients) may exceed the strength of one of the phases.

Organofunctional silanes are hybrid organic–inorganic compounds that are used as coupling agents across the organic–inorganic interface to help overcome these two obstacles to composite performance. One of their functions is to use the hydrolytic action of water under equilibrium conditions to relieve thermally induced stresses across the interface. If equilibrium conditions can be maintained, the two problems act to cancel each other out.

Coupling agents are defined primarily as materials that improve the practical adhesive bond of polymer to mineral. This may involve an increase in true adhesion, but it may also involve improved wetting, rheology, and other handling properties. The coupling agent may also modify the interphase region to strengthen the organic and inorganic boundary layers.

Primary suppliers of coupling agents have been cooperative in describing the composition and chemistry of their products in order to help formulators to use them intelligently. Commercial formulated sizes, primers, and adhesives, however, are generally considered proprietary and are described only in terms of performance. This book discusses applications of silane coupling agents in areas where full disclosure of chemistry and composition are possible, with essentially no mention of proprietary formulations.

Fiberglass-reinforced plastic composites have been developed to high-performance materials since their introduction in 1940. For the first 20 years, the development was based on practical performance tests, and concepts of mechanisms involved were based largely on indirect evidence. In recent years, the interface has been studied extensively by advanced analytical techniques of surface science. This book covers both areas of study into the early 1980s. It combines practical technology with fundamental science to arrive at a useful understanding of the interface in composites. It should help scientists design experiments to further their studies of the interface, and should help technologists develop better composites from available materials.

It is now possible to obtain silane coupling agents to modify common mineral surfaces for reinforcement of almost any organic polymer.

Even the best coupling agent for a given composite may perform poorly if it is not applied properly. Orientation of the molecules and physical properties of the film (solubility, fusibility, and mechanical properties) can be controlled by the method of application, and may be as important as the chemistry of the selected silane.

An understanding of materials and mechanisms of adhesion should allow consistent preparation of composites that utilize the full capabilities of fillers, reinforcements, and matrix resins.

Edwin P. Plueddemann

# Contents

Chapter 4.  SURFACE CHEMISTRY OF SILANES AT
           THE INTERFACE

Chapter 5.  NATURE OF ADHESION THROUGH SILANE
           COUPLING AGENTS

Chapter 6.  PERFORMANCE OF SILANE COUPLING
           AGENTS

Chapter 7.  PARTICULATE-FILLED COMPOSITES

Chapter 8. OTHER APPLICATIONS OF SILANE COUPLING
AGENTS

# Silane Coupling Agents

SECOND EDITION

# 1 | General Concepts

## 1.1. History of Coupling Agents

Although bonding of organic polymers to inorganic surfaces has long been a familiar operation (e.g., protective coatings on metals), a major need for new bonding techniques arose in 1940 when glass fibers were first used as reinforcement in organic resins. The specific strength-to-weight ratios of early glass–resin composites were higher than those of aluminum or steel, but they lost much of their strength during prolonged exposure to moisture. This loss in strength was attributed to the debonding of resin from hydrophilic glass by the intrusion of water.

Careful preparation of glass–resin bonds under dry conditions did not produce more water-resistant composites. Even bonds of resin to fresh glass surfaces, fractured in liquid resin before polymerization, did not withstand the action of water. Clearly, the interface between such dissimilar materials as an organic polymer and a mineral did not allow the formation of a water-resistant bond.

The concept that two dissimilar materials may be held together by a third intermediate material as coupling agent was used by Plato[1] to explain how a universe made up of four elements—earth, air, fire, and water—could exist as a homogeneous whole:

> It is not possible for two things to be fairly united without a third, for they need a bond between them which shall join them both, that as the first is to the middle, so is the middle to the last, then since the middle becomes the first and the last, and the last and the first both become middle, of necessity, all will come to be the same, and being the same with one another, all will be a unity.

Because organofunctional silicones are hybrids of silica and of organic materials related to resins, it is not surprising that they were tested as coupling agents to improve bonding of organic resins to mineral surfaces.

Modification of the interface between an organic polymer and an inorganic substrate may have many beneficial results in composite manufacture, but "coupling agents" should be defined as materials that improve the chemical resistance (especially to water) of the bond across the interface. Although any polar functional group in a polymer may contribute to improved adhesion to mineral surfaces, a methylacrylate–chrome complex (Volan A®)[2] and the various organofunctional silanes have shown the most promise as true coupling agents.

In 1947, Ralph K. Witt et al. of Johns Hopkins University, in a "confidential" report to the Navy Bureau of Ordnance, observed that allyltriethoxysilane on glass fibers gave polyester composites with twice the strength of those where the glass was treated with ethyltrichlorosilane.

The Bjorksten organization was given a contract by the U.S. Air Force in about 1949 (AFTR 6220) to explore the effect of glass fabric treatments on polyester laminate wet strength properties. A total of 2000 compounds was screened. The best of these, and still good by today's standards, was a nonaqueous solvent treatment (BJY) based on an equimolar adduct of vinyl trichlorosilane and $\beta$-chloroallyl alcohol.[3]

Polyester laminate strength data obtained by the Bjorksten group using their BJY treatment are shown in Figure 1.1. Laminates with treatments 114 (chrome) and 112 (no finish) are also shown for comparison. After 5 hr

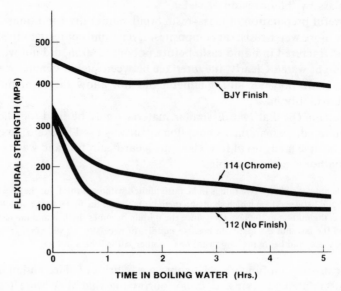

**Figure 1.1.** Polyester laminate strengths in 1950 from data of Bjorksten.[3]

in boiling water, the flexural strength of the BJY laminate was still above 400 MPa and was, in fact, substantially greater than the original dry strength of the other laminates.

Over a hundred different organofunctional silanes were evaluated as coupling agents in glass-reinforced polyester and epoxy composites by Plueddemann et al.[4] in 1962. In general, the effectiveness of the silane as a coupling agent paralleled the reactivity of its organofunctional group with the resin. Styrene-diluted polyester resins are complex mixtures of oligomeric maleates and fumarates in a reactive monomer, but the performance of silanes showed good correlation with their expected relative reactivity in copolymerizing with styrene (Table 1.1).

Two events in 1963 had an impact on subsequent interface investigations. The first event was the formation of the Ad Hoc Committee on the Interface Problem in Fibrous Composites by the Materials Advisory Board. This committee was charged with investigating the problem in depth in order to define problem areas and to come up with recommendations on what sort of research should be sponsored. Universities, government, and industry were represented by persons from many disciplines. the committee report (MAB-214-M) was published in 1965. A most important finding was that the principles of surface chemistry related to the problem had been virtually ignored for about 15 years. A major recommendation was that research efforts be scaled up to a level of 50 senior investigators for a period of 5 years.

The second event to have an impact on the type of research done was a 2-day meeting by invitation only on the interface problem at the Naval

Table 1.1. Relation of Reactivity with Styrene to
Effectiveness of Silane Finishes in Polyester Composites

| Silane finish reactive group | Relative reactivity with styrene | Order of wet strength of polyester laminates |
|---|---|---|
| Fumarates | 3.3 | 2 |
| Methacrylates | 1.9 | 1 |
| Acrylates | 1.34 | 3 |
| Isoprene | 1.3 | 9 |
| Styrene | 1.0 | 5 |
| Vinyl sulfides | 0.2 | 6 |
| Maleates | 0.15 | 4 |
| Crotonates | 0.05 | 11 |
| Vinyl Si | 0.05 | 7 |
| Allyl Si | 0.03 | 8 |
| Allyl esters | 0.011 | 10 |
| Vinyl ethers | 0.01 | 12 |

Research Laboratory in Washington in June 1963. This was attended by about 75 persons from government, universities, and industry. The proceedings were not published, but this meeting had an impact on the type of interface research that the government sponsored in succeeding years. Surface chemistry investigations began in earnest exploring basic phenomena at the interface. The Air Force followed by sponsoring interface research using radioisotope methods.

Instrumental techniques have been developed in recent years to study the chemistry of the interface in reinforced composites through laser Raman spectroscopy and Fourier transform infrared spectroscopy. It is now possible to observe chemical reactions at the interface that had previously been inferred from indirect evidence[5] (Chapter 4).

A consistent forum for discussion of interface phenomena have been the "interface" sessions held for over 40 years at the Annual Technical Conferences of the Reinforced Plastics Division of the Society of Plastics Industry (SPI). In recent years, many technical societies have organized symposia on "the interface." The International Conference on Composite Interface was first held at Case-Western Reserve University in 1986 and now is established as a biannual meeting devoted entirely to composite interface studies.

## 1.2. Definitions

Organofunctional silanes may be used as adhesion promoters between organic polymers and mineral substrates under a variety of circumstances. The silane adhesion promoter, or "coupling agent," may function as (1) a finish or surface modifier, (2) a primer or size, or (3) an adhesive, depending on the thickness of the bonding material at the interface.

A surface modifier or "finish" may theoretically be only a monomolecular layer, but in practice it may be several monolayers thick. The material functions to chemically modify a surface without contributing any mechanical film properties of its own.

A primer or size is generally 0.1 to 10 $\mu$m thick, and must have adequate mechanical film properties, such as rigidity, tensile strength, and toughness to carry the mechanical load when the composite is stressed. The primer may be a layer of hydrolyzed silane or a silane-modified polymer. It is generally applied from a solvent to ensure uniform coverage of the surface.

An adhesive is a gap-filling polymer used to bond solid adherents such as metals, ceramics, or wood, whose solid surfaces cannot conform to one another on contact. Pure silanes are rarely used as adhesives; rather, they are used to modify gap-filling polymers or polymer precursors to improve surface adhesion. Silane monomers may be used in integral blends of fillers and liquid resins in the preparation of composites. The modified polymer "adhesive" in this case is termed a matrix resin.

The fiberglass industry uses terminology derived from the textile industry. Much of the glass fiber made for a filament winding or as chopped roving is given a single and final treatment at the time of forming. Such a treatment is correctly called a size. The treating solution is a mixture of many things, the more important being lubricant, antistat, binder, and coupling agent. The first of these provides surface lubricity, which prevents abrasive damage during handling. A binder is necessary for strand integrity, because the single filaments do not normally pack well due to static electricity. There may also be other factors. (An antistat is added to prevent buildup of static electricity during mechanical handling.) The glass may pick up 1% or 2% of the size, of which about 10% will be silane adhesion promoter. Size formulations on commercial glass are proprietary for each manufacturer and are formulated to combine good handling properties of the roving with good adhesion (primer action) for particular resins. Different sizes are formulated for polyesters, epoxies, etc.

The other type of treatment is called a finish. Glass fibers that are to be given a finish are also given a size at the time of forming. This size contains ingredients that provide surface lubricity and binding action, but usually no coupling agent is present. These fibers in the sized condition can be plied and woven into fabric without any significant damage to the fibers. Before the finish is applied it is necessary to remove this size. This is accomplished by heating the fabric (or roving) in air-circulating ovens to burn the size away. The finish is then applied from solution to the heat-cleaned fabric to complete the treatment. The major chemical in the finish treatment is a coupling agent, but other materials may be included to improve lubricity, resin wet-out, etc. The total pickup of finish on glass fibers is generally about 0.1%. Each weaver supplies glass with proprietary finishes for reinforcement of specific resins. Although modification of silanes with organic polymers may provide some improvement in primer performance where film properties are important, a silane solution with only minor modification appears to be the best finish on glass cloth.

Silane coupling agents for reinforced composites are required to fulfill rather complex responses, which may vary with different methods of application. For treatment of fiberglass it is required that the silane be soluble in water, and that dilute aqueous solutions remain predominantly monomeric for at least one day in the treating bath. When dried on the glass surface, the coupling agent must condense to polysiloxane structures that retain a degree of solubility in order to be compatible with the matrix resin. During the curing of the composite it is advantageous if the interphase region controlled by the coupling agent is hydrophobic and highly cross-linked. Since fiberglass may be stored for months or years, it is important that the siloxane structure does not cross-link too highly during storage.

When formulated as primers, silanes must be stable in more concentrated solutions of suitable solvents such as aqueous alcohols. A minimum shelf stability of 6 months is desired. After filming on a surface, the condensed siloxanes need to be soluble and fusible for only several minutes or hours. The dried primer film must be compatible with the matrix resin and contribute a high degree of cross-linking at the interface.

## 1.3.  Evaluation

Although coupling agents may perform several useful functions at the interface in mineral-reinforced composites, the coupling agent is expected first of all to improve the adhesion between resin and mineral, and then to improve retention of properties in the presence of moisture. Several simple laboratory tests are recommended to evaluate any new potential coupling agent. More extensive and sophisticated evaluation then may be used to refine the application for optimum performance.

### 1.3.1.  Thin Film Tests

A simple accelerated test for adhesion that is useful for rapid screening of adhesion promoters is to form a thin film of polymer on primed glass microscope slides or metal coupons and soak in water until the film can be loosened with a razor blade. There is usually a clear transition between cohesive failure and interfacial failure. Since diffusion of water to the interface is very rapid in this test, the time to failure depends on the interfacial properties and may differ several-1000-fold between unmodified polymers and silane-modified polymers. Different silane primers may differ several-100-fold in performance with a given polymer.[6]

Commercial microscope slides provide convenient clean surfaces for observing adhesion of resins to glass. The slides may be dipped in dilute (e.g., 0.5%) aqueous solutions of silane or wiped with a 10–25% solution of prehydrolyzed silane in an alcohol (primer). Thermoplastic resins are then fused under light pressure against the treeated slides. Thermosetting resins are applied as liquids and cured against the treated slides. The finished films on glass are observed for initial adhesion and then soaked in water at room temperature or at elevated temperature to determine resistance of the bond to intrusion by water.

Adhesion of elastomeric polymers may be measured directly as peel strength. Rigid polymers cannot be peeled. Attempts to lift the film will indicate a qualitative degree of adhesion up to cohesive failure in the polymer. When thick rigid films with good adhesion to glass are immersed in water, the film may suddenly pop off the slide but carry a complete layer of glass with it. Adhesion provided by the silane may be rated as the length of time a film may be soaked in water (assuming there was good dry adhesion) before there is loss of adhesion. In general, there is very good correlation between the adhesion observed in microscope slide tests and the performance a silane will provide in mineral filled elastomers[7] (Figure 1.2) or resins.[8]

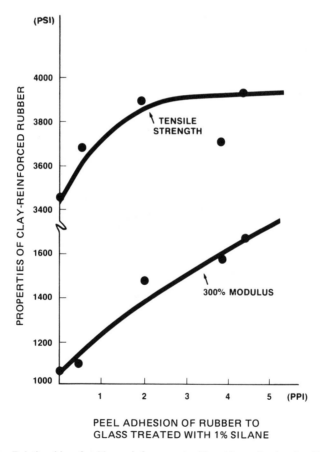

PEEL ADHESION OF RUBBER TO
GLASS TREATED WITH 1% SILANE

**Figure 1.2.** Relationship of rubber reinforcement with rubber adhesion to silane-treated surfaces.

### 1.3.2. Castings

Silane coupling agents for particulate fillers in liquid thermosetting resins are most readily evaluated in castings. The filler may be pretreated with silane solutions, or the silane monomer may be added as an integral blend with the filled resin. Different levels of a given silane may be compared with different fillers[9] (Figure 1.3), or different coupling agents may be compared with a given resin filler mix. Silane additives may have surface-active properties, as shown by the reduction in viscosity, but all surface-active additives (e.g., titanates) do not act as coupling agents, as shown by retention of wet strength (Table 1.2).

A convenient technique for making castings is to mix sufficient filler with resin to give a pourable viscosity. The filler may be pretreated with silane, or 0.3–0.5% (based on filler) or silane monomer may be added as integral blend during mixing. The mixture is de-aired under vacuum and cast into 8 × 75 mm test tubes pretreated with a release agent. The castings are cured and evaluated by measuring flexural strength wth three-point loading at a 2-in. span.

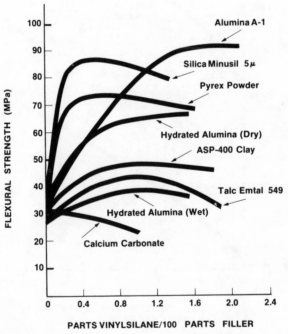

**Figure 1.3.** Flexural strength as a function of parts vinylsilane per 100 parts filler (Buton® casting).

Table 1.2. Polyester Castings with 50% Silica[a]

| Additive (0.3%) based on filler | Viscosity of mix, cp | Flexural strength, MPa | |
| --- | --- | --- | --- |
| | | Dry | 2-hr water boil |
| None | 24,500 | 115 | 70 |
| Z-6030[b] | 22,000 | 163 | 143 |
| Z-6032[c] | 8,700 | 156 | 139 |
| TTM-33[d] | 10,000 | 135 | 72 |

[a] Minusil 5 $\mu$m in Rohm & Haas Paraplex P-43.
[b] Dow Corning® methacryloxypropyltrimethoxysilane.
[c] Dow Corning® vinylbenzyl cationic silane.
[d] Kenrich Petrochemical's isopropyltrimethacroyltitanate.

### 1.3.3. Pultrusion Rods

Although silane coupling agents may be tested on microscope slides or in castings with glass microbeads, a rapid test method with glass fibers was desired, because surface composition and morphology of glass fibers are quite different from those of microscope slides and microbeads.

Biefield and Philips described a simple pultrusion test in 1952. Unsized (or heat-cleaned) glass roving was treated with the desired coupling agents and fabricated into pultrusion rods. Glass tubing treated with a release agent is used as pultrusion form. A string or fine wire is tied around the middle of sufficient glass roving so the doubled strands fill the form tightly. The roving is immersed in liquid resin and pulled with the string into the end of the tube immersed in resin. The bottom of the tube is then sealed and the resin cured. The composite is pulled from the tube and cut into lengths for testing. In evaluating epoxy resins, a length of Teflon® tubing in a slightly larger glass tubing (to keep it straight) may be used as a form for easy release of resin. The pultruded composites show very high flexural strengths because of the parallel orientation of glass fibers.

Similar techniques have been used in evaluating composites of more exotic fibers such as graphite, Kevlar®, and alkali-glass fibers.

Flexural strengths of some typical pultruded polyester rods prepared with silane-treated E-glass rovings are shown in Table 1.3. Much higher strengths were observed than in composites containing particulate silica (Table 1.2), but the order of improvement of wet-strength retention provided by silane coupling agents was similar in both series.

### 1.3.4. Woven Glass-Reinforced Laminates

Silane finishes on glass cloth are commonly evaluated in thermosetting resin laminates. Glass with a good coupling agent generally produces

**Table 1.3.** Polyester Pultrusion Rods, 70% Glass

| Silane treatment on glass | Flexural strength, MPa | |
|---|---|---|
| | Dry | 4-hr boil |
| None | 966 | 414 |
| 0.05% Z-6030[a] | 1072 | 877 |
| 0.2% Z-6030 | 1146 | 963 |
| 0.05% Z-6032[b] | 1002 | 886 |
| 0.2% Z-6032 | 1133 | 986 |

[a] Dow Corning® methacryloxypropyltrimethoxysilane.
[b] Dow Corning® vinylbenzyl cationic silane.

laminates with good clarity and little fiber show. When laminates are boiled in water the intrusion of water into the resin–glass interface may often be observed by a whitening of the laminate, the appearance of fiber weave, and a roughening of the surface. The relative performance of a series of coupling agents on glass can often be estimated by visual examination of laminates after boiling in water from 2 to 72 hr.

Heat-cleaned woven glass cloth obtained from commercial weavers may be dipped in 0.1% aqueous silane solutions and dried. Considerable art is involved in preparing polyester or epoxy laminates for evaluation. Fabric, for example, is impregnated with an acetone solution of epoxy resin, dicyandiamide, and benzyldimethylamine and dried to obtain prepregs that are pressed at 69 MPa (1000 psi) at 165°C for 60 min to obtain composites with a resin content of about 38%.

Epoxy laminates for circuit boards are designed to maintain insulation values after repeated humidity cycling. Because electrical or mechanical breakdown involves the intrusion of water into the resin–glass interface, it is possible to estimate the electrical performance of laminates from simple mechanical strength measurements after prolonged boiling in water. A comparison of epoxy laminates prepared with various coupling agents on glass is shown in Table 1.4.[10] The various silanes are clearly superior to a chrome-complex coupling agent, and, among the silanes, the vinylbenzyl-cationic-functional silane imparts the best water resistance.

The deleterious effect of aqueous attack of the interface can be accelerated by boiling the test coupons in commercial pressure cookers or autoclaves. Entry of water into the laminate can be observed by testing insulation resistance or by immersing the boiled coupon into molten solder. Sudden conversion of water in the interior of the laminate to steam causes delamination, which is readily observed visually.

**Table 1.4.** Mechanical Strengths of Epoxy Laminates (MPa),
(G-10 formulation with style 7678 glass cloth)

| Finish on glass | Initial | | 2-hr boil | | 200-hr boil | |
|---|---|---|---|---|---|---|
| | Flex | Comp | Flex | Comp | Flex | Comp |
| None | 448 | 290 | — | — | — | — |
| Chrome complex (Volan® A) | 503 | 345 | 421 | 282 | 227 | 138 |
| ≡SiCH$_2$CH$_2$CH$_2$Cl | 586 | 324 | 503 | 310 | 345 | 241 |
| ≡SiCH$_2$CH$_2$CH$_2$NH$_2$ · HCl | 544 | 338 | 482 | 303 | 353 | 234 |
| ≡Si(CH$_2$)$_3$NHCH$_2$CH$_2$NHCH$_2$C$_6$H$_5$ · HCl | 586 | 365 | 455 | 338 | 324 | 320 |
| ≡Si(CH$_2$)$_3$NHCH$_2$CH$_2$NHCH$_2$C$_6$H$_4$—CH = CH$_2$ | 586 | 358 | 550 | 310 | 413 | 310 |
| ≡(CH$_2$)$_3$OCH$_2$$\overset{O}{\overset{/\backslash}{CH}}$CH$_2$ | 509 | 339 | 457 | 332 | 322 | 221 |

There is, as yet, no standard "pressure cooker test," although many test laboratories have used some modification of this technique to separate "good" laminates from "bad" laminates. Obviously, a test can be made so severe that ultimately all laminates will fail. It may be that a pressure cooker test of a laminate should not exceed the glass transition temperature of the cured epoxy resin for a fair evaluation.

### 1.3.5. Injection-Molded Thermoplastics

Adhesion of polypropylene fused to glass at 200°C was compared with properties of filled injection-molded polypropylene with comparable surface treatments for adhesion. When 0.5% vinylbenzyl cationic silane (Dow Corning® Z-6032) was used as a standard treatment on fillers such as glass microbeads, Wollastonite, mica, and talc, it was observed that mechanical properties (strength, modulus, and heat distortion temperature) increased with increasing adhesion observed on primed microscope slides (Table 1.5). Addition of 1% dicumylperoxide to the silane improved its performance as adhesion promoter.[11]

The full value of silane coupling agents on fillers in injection-molded thermoplastics has not been realized because the high shear across the interface destroys bonds to the coupling agent during fabrication. A new coupling method has been proposed in which an acid-functional silane on

Table 1.5. 35% Wollastonite[a]-Filled Polypropylene Compared with
Microscope Slide Adhesion

| Treatment on surface | | Additive[c] to polymer | Adhesion to glass | Flexural strength of molding, MPa |
|---|---|---|---|---|
| Silane[b] | Peroxide | | | |
| No | No | No | None | 60 |
| Yes | No | No | Poor | 64 |
| Yes | No | Yes | Good | 67 |
| Yes | Yes | No | Good | 70 |
| Yes | Yes | Yes | Excellent | 79 |

[a] Nyad-400® (Nyco Div. of Processed Minerals, Inc.).
[b] Dow Corning® Z-6032 vinylbenzyl cationic silane.
[c] 1% Chlorez-700® (Cover Chem.) based on polymer.

the filler bonds to acid-functional polymers in the presence of cations through ionomer bonds. Ionomers are fluid under conditions of high temperature and high shear, but reform at room temperature to tough water-resistant structures (see Section 7.5.5).[11]

### 1.3.6. Single-Fiber Tests

The measurement of adhesion is a complex problem. Values observed in a debonding experiment depend strongly on the mode of failure. Numerous tests have been devised to measure the stress-transfer capability between resins and single reinforcing fibers. Although these may not be measurements of true shear strength, they are informative in showing the effect of resin properties and fiber finishes on resistance to debonding.

A microbonding test on cross sections of real composites was described by a group headed by Mandel et al.[12] A diamond-tipped probe was pushed against a single fiber end under an optical microscope to the first optical evidence of debonding. The probe was instrumented with an extensometer, which served as a load transducer. The load at debond was used to calculate interfacial debonding strength by a finite-element program. Operation of this device is entirely manual, and reproducible results can be obtained only by a highly skilled and trained operator.

The McGarry–Mandel test has been automated and computerized by Dow scientists in the form of an interfacial testing system that is available commercially from Schares Instrument Corporation (Houston, TX). Caldwell and Jarvie demonstrated the use of this instrument in studying the interfacial strength of carbon fiber/polymer composites.[13] Comparing two types of graphite and various modifications of Dow XU 7177.02 cyanate ester resin, they found that edge delamination strength has an inverse correlation with interfacial shear strength as measured by a short-beam shear test (Figure 1.4). Composite performance is, therefore, a complex combination of adhesion, debonding, and other forms of energy absorption in the interface region.

In an ingenious test described by Fraser et al.[14] a single fiber encapsulated in a tensile test specimen is subjected to traction to an elongation greater than that required to fracture the fiber. (The matrix resin must have elongation-to-break significantly greater than that of the fiber.)

When a specimen containing a single fiber is subjected to tensile deformation and the plastic matrix deforms beyond the elongation-to-break of the fiber, the fiber will break at its weakest point. One now has two pieces of fiber of different lengths. Under continued deformation, the fiber stress continues to increase until it reaches the breaking stress of one of the two fragments, at which point a second fracture occurs. This process will continue until all the fragments are smaller than their critical lengths, after which time fiber fracture is no longer possible. If the tensile strength of the

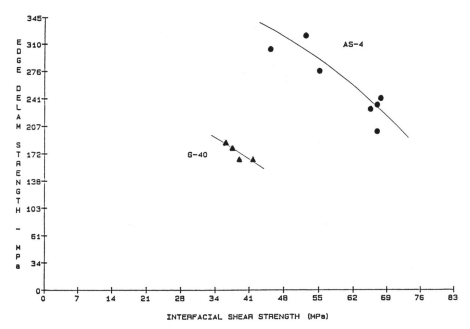

**Figure 1.4.** Edge delamination strength vs. interfacial shear strength in graphite fiber composites.[13]

fiber is known and is independent of fiber length, one may calculate an effective interfacial shear strength from the measured critical length.

Single fibers of E-glass with an aminofunctional silane size were pulled in Nylon 6, polypropylene, and a carboxyl-modified polypropylene. The cumulative distributions of fragments from ten specimens of each system indicated an effective interfacial strength of 34.4 MPa for Nylon 6, 7.1 MPa for polypropylene, and 21.0 MPa for the carboxy-modified polypropylene.

Drzal et al.[15] reported similar studies on graphite fibers embedded in m-PDA-cured epoxy resin. A commercial surface-treated fiber was compared with a similar fiber without surface treatment. Interfacial shear strengths calculated from the observed distributions of fiber fragments were 40.8 MPa for treated fiber and 15.6 MPa for untreated fiber.

### 1.3.7.  Other Tests

The effect of adhesion at the glass–resin interface on impact strength of composite was studied by comparing short-beam shear tests of glass-fiber-reinforced epoxies and polyesters with cantilever cleavage tests and Charpy impact tests on similar composites.[16] Initiation energy-of-fracture increased with increasing interlaminar shear strength. The propagation energy displayed a minimum at a critical value of interlaminar shear strength. The highest total impact strength resulted from a release agent on the reinforcement. Somewhat better adhesion obtained with clean glass gave a minimum impact strength, which increased again with good coupling agents on the glass surface (Figure 1.5). Improved impact strength, therefore, cannot be used as indication of coupling at the interface, because it could also indicate abhesion or release properties.

Electrical insulation tests during humidity cycling are necessary for proper evaluation of glass-reinforced circuit boards, but such specialized tests are not recommended for initial screening of coupling agents.

### 1.3.8.  Test for Water Resistance of Bond

No perfect correlation has been established between any accelerated environmental tests and performance of a composite in outdoor weathering, because outdoor weathering is so highly variable.

A 2-hr water boil is considered to be about equivalent to 30 days immersion in room-temperature water.[17] As the temperature of boiling water is raised to 125°C (pressure cooker), the rate of deterioration of the bond increases by another factor of 10.[18] In one series of tests, the effect produced by 60 cycles in a dishwasher was equivalent to a year of outdoor weathering on surface appearance of reinforced polyesters.[19]

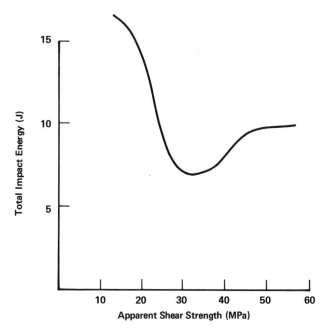

**Figure 1.5.** Impact energy vs. short-beam shear strength for fiberglass-reinforced polyester laminates.[16]

## 1.4. Nonsilane Coupling Agents

### 1.4.1. Chromium

The only nonsilane materials that have shown significant coupling activity are the chromium complexes supplied as Volan® by duPont. Volans® are described as coordination complexes of carboxylic acids with 2 mol of trivalent chromium chlorides.[20] In water, the chromium chlorides hydrolyze to basic salts. These basic salts form oligomeric salts through "olation" of hydroxyl groups on adjacent chromium atoms. The hydroxyl groups also bond to silanol groups of a glass surface through hydrogen bonding and possibly covalent oxane bonds. The organic acid group develops a fairly stable bond to chromium by being coordinated to adjacent chromium atoms (cf. Table 6.1.). The chromium complex with methacrylic acid was used for many years as a standard finish on glass for reinforcement of polyesters and epoxies. The methacrylic function could be homopolymerized and was believed to copolymerize with the polyester resin. The mechanism for coupling to epoxy resins is less clear, although it is generally observed that reactive double bonds coreact in some way in epoxy curing reactions.

Chromium complexes of fumaric acid (Volan® 82) have been reported to be very effective in bonding aluminum to polyethylene.[21] Volan® finishes on glass also contributed good antistatic and other handling properties to glass fibers so that the green color contributed by chromium was recognized universally as a sign of "coupled" glass for reinforcement of plastics.

As specifications for water resistance of reinforced plastics were made more stringent, the chromium complexes were not able to match the perform-ance of newer silane coupling agents. Although the acid complexes had unusual stability in water, the bond to chromium could not compare with the stable carbon–silicon bond of organofunctional silanes.

### 1.4.2. Orthosilicates

Very good primers may be formulated from orthosilicate esters and metal orthoesters for bonding organic polymers to glass, metals, and even plastic surfaces. As will be shown in Chapter 3, the equilibrium constant of hydrolysis of orthosilicates to metals and glass are even more favorable than those of organofunctional silane coupling agents. These primers func-tion by partial hydrolysis to oligomers that interdiffuse with liquid polymers and then cross-link to form interpenetrating polymer networks at the inter-face. The primer ultimately cross-links to silica, which has outstanding heat stability and hydrolytic stability. The major limitation of orthosilicate primers is the rather narrow "window" of time during which the films are soluble and compatible with matrix polymers. The polymers must be applied within a few minutes or hours, depending on the ambient temperature and humidity. Such primers are not suitable as coupling agents for pretreatment of fillers or fiberglass because these are often stored for months before they are incorporated into resins. The reaction product of $SiCl_4$ and glycidyl methacrylate[22] applied to glass from organic solvents produced very credi-table polyester laminates with good resistance to 2-hr water boil (Table 1.6):

$$SiCl_4 + CH_2{=}CCH_3COOCH_2\overline{CHCH_2O} \rightarrow$$

$$Cl_3SiOCH_2CHClCH_2OCOC(CH_3)CH{=}CH_2$$

A methylcrylate-functional alkoxysilane obtained by alcoholysis of $(MeO)_4Si$ with hydroxyethylmethacrylate showed no value as a coupling agent[23]:

$$CH_2{=}C(CH_3)COOCH_2CH_2OH + (MeO)_4Si \rightarrow$$

$$MeOH + CH_2{=}C(CH_3)COOCH_2CH_2OSi(OMe)_3$$

**Table 1.6.** Polyester Glass Laminates
(with 2% methacrylate monomer)

| Functional group of methacrylate additive | Flexural strength of composite, MPa | |
|---|---|---|
| | Dry | Wet, 2-hr boil |
| No additive | 386 | 240 |
| $-COOH$ | 457 | 296 |
| $-CH_2CH_2OH$ | 440 | 285 |
| $-CONH_2$ | 408 | 268 |
| $-CH_2\overline{CH_2CH_2O}$ | 408 | 281 |
| $-CH_2CH_2(OH)CH_2OPO(OH)_2$ | 438 | 307 |
| $-(CH_2)_3Si(OMe)_3$ | 633 | 588 |
| $-$Chrome complex[a] | 502 | 427 |
| $-CH_2CH_2ClCH_2OSiCl_3$[b] | 622 | 539 |
| $-CH_2CH_2OSi(OMe)_3$[b] | 347 | 232 |

[a] DuPont Volan® preapplied to glass from water.
[b] Applied to glass from toluene.

Reaction products of $SiCl_4$ and allylglycidyl ether and propylene oxide or other epoxy monomers were described as complex compositions in aqueous sizes for treating fiberglass roving for reinforcement of polyester composites.[24] Both initial flexural strengths and strengths after 4 hr in boiling water were comparable to those of polyester composites prepared from glass roving using a methacrylate-organofunctional silane in the size formulation.

### 1.4.3. Other Orthoesters

Various inorganic esters have been claimed as coupling agents for reinforced plastics, including aminobenzyl phosphonates,[25] dicetyliso-propylborate,[26] alkoxy compounds of aluminum, zirconium, and titanium,[27] zircoaluminates,[28] and numerous substituted titanates.[29] Improvement in dispersion of fillers in polymer and improved properties of composites are sometimes described as a "coupling effect," although the action of these polar esters may be more nearly like that of hydrophobic wetting agents used in the paint industry for improved dispersion of pigments.[30]

Certain inorganic ester primers are very effective in bonding epoxies to anodized aluminum even though they appear to completely cover the microporosity of the anodized surface.[27] It is believed that these primers deposited an amorphous oxide layer that was well-bonded to the aluminum, and retained an optimum arrangement of reactive hydroxyl groups for bonding to the polymer.

### 1.4.4.  Other Organofunctional Compounds

Polar organofunctional groups in a polymer generally improve the adhesion and water-resistance of bonds to hydrophilic mineral surfaces. Polar functional monomers added to a thermosetting resin or preapplied to reinforcement provide some improvement in the composite. The difference between a coupling agent and a polar additive may be a matter of degree. A silane coupling agent is much more effective than any polar organic monomer in improving adhesion, and the degree of improvement obtainable with a small amount of silane is greater than could be obtained with any amount of polar monomer included in the composite. This relationship was shown by comparing a number of organofunctional methacrylate monomers as additives in polyester resins laminated with heat-cleaned glass cloth (Table 1.6).[23]

## 1.5.  Theories of Bonding Through Coupling Agents

Because minute proportions of coupling agents at the interface have such a profound effect on the performance of composites, it is expected that elucidation of the coupling mechanism will be helpful in understanding the fundamental nature of adhesion of organic polymers to mineral surfaces. Consideration of the various theories of adhesion may be useful in understanding the mechanism of adhesion through silane coupling agents, but even more importantly, an understanding of the mechanism of coupling through silanes should clarify the general concept of adhesion—especially adhesion of organic polymers to hydrophilic inorganic surfaces.

### 1.5.1.  The Chemical Bonding Theory

The chemical bonding theory is the oldest, and is still the best known, of the theories.[4] Coupling agents contain chemical functional groups that can react with silanol groups on glass. Attachment to the glass can thus be made by covalent bonds. In addition, coupling agents contain at least one other, different functional group which could coreact with the laminating resin during cure. Assuming that all this occurs, the coupling agent may act as a bridge to bond the glass to the resin with a chain of primary bonds. This could be expected to lead to the strongest interfacial bond.

Indirect evidence for coreaction of the organofunctional group with thermosetting resins was easy to obtain. In terms of strength properties, vinyltrichlorosilane-finished glass gives polyester laminates with dry and wet strengths about 60% greater than those of laminates of ethyltrichlorosilane-finished glass. Similarly, allyltrichlorosilane gives about 70% greater values than the saturated derivative, propyltrichlorosilane. Glass finished with the isobutyric acid complex of chromium gave laminates whose strength after boiling was almost exactly halfway between the strength of control laminates containing no complex and those made with the methacrylic acid complex. The two compounds in each pair differ chemically only by olefinic unsaturation. On the basis of the theory, an unsaturated polyester laminating resin should be able to copolymerize with olefinic groups in a finish (Table 1.7).

In comparing an extensive series of unsaturated-functional trimethoxysilanes as coupling agents for polyester resins, it was observed (Table 1.1) that effectiveness of the silane as a coupling agent was directly related to the reactivity of the functional group in copolymerizing with styrene-diluted resin. Thus, a methacrylate-ester-functional silane was a very good coupling agent, while the isomeric crotonate-ester-functional silane was a very poor coupling agent.

With the development of Fourier transform infrared (FT-IR) spectroscopy, it was possible to observe chemical reaction in the silane interphase region of silane-treated glass fibers in a styrene-diluted polyester during cure.[31] First, it was observed that E-glass fibers inhibit the polymerization of polyester as was proposed from observations of cure exotherms.[32] A methacrylate-functional silane formed a multilayer on an E-glass fiber and polymerized completely during resin cure, while a vinylsilane layer did not react completely. Furthermore, the carbonyl frequency of the polymerized coupling agent differed from that of homopolymerized material, indicating

**Table 1.7.** Polyester Laminates with Various Treatments on Glass

| Treatment on glass | Flexural strength, MPa | |
| --- | --- | --- |
| | Dry | Wet, 2-hr boil |
| None (heat-clean) | 386 | 240 |
| Ethyltrichlorosilane | 238 | 179 |
| Vinyltrichlorosilane | 496 | 406 |
| Propyltrichlorosilane | 236 | 183 |
| Allyltrichlorosilane | 399 | 402 |
| i-Butyric-Cr complex | — | 310 |
| Methacrylic-Cr complex | 502 | 427 |

that true copolymerization had taken place. An important additional observation was that the methacrylate group of the coupling agent was very sensitive to drying conditions. Much of the methacrylate unsaturation was lost in 1 hour at temperatures above 100°C in air. Care, therefore, must be used in drying methacrylate-functional silanes on reinforcement surfaces to obtain full benefit of the coupling agent.

Covalent bonding between coupling agent and mineral surface was somewhat more difficult to demonstrate. Silanols of a hydrolyzed silane coupling agent will first form hydrogen bonds with hydroxyl groups of a mineral surface or possibly with a layer of water on the mineral surface. The question arose whether those silanol groups ever condense to oxane bonds with the surface, or whether such bonds are necessary for the coupling mechanism to operate.

A methacrylate-functional silane applied to glass from water and dried under mild conditions had considerable mobility on the glass surface, as shown by its ready displacement by acetoxypropyl silanols. Still, glass treated with methacrylate-functional silane and dried at room temperature gave very good polyester laminates.[33] Either oxane condensation with the glass was not necessary or such bonding took place during the composite curing cycle. Similar performance is expected when silane monomers are added as integral blends to filled polymer systems before cure.

In comparing the properties of several polyester laminates prepared with the methacrylate-functional silane as coupling agent, it was observed that drying treated glass briefly at 110°C gave much better performance than drying at room temperature. Performance fell off, however, when treated glass was dried at 160°C. When silane monomer was used as integral additive, better performance resulted from curing at higher temperatures. After the laminate was cured, an additional postcure at still higher temperatures was of no benefit (Table 1.8).

When glass was treated with silane coupling agents in organic solvents, it was shown that addition of an amine catalyst increased the degree of condensation of silane with the glass surface (as shown by infrared absorbance of hydroxyls) and improved the properties of polyester composites prepared from treated glass microbeads (Table 1.9).[34]

Infrared spectroscopy has given direct evidence of silanol condensation between coupling agent and high-surface-area silica, but glass fibers have too low surface area to show such reactions by infrared spectroscopy. The greater sensitivity of FT-IR spectroscopy has made possible a direct study of silanol reactions with E-glass fiber. It was concluded that coupling agent near the glass surface has a high degree of molecular orientation, and that silanol condensation is enhanced by the glass surface. Cross-condensation between coupling agent and glass was observed during drying.[35]

**Table 1.8.** Polyester–Glass Laminates with Methacrylate Silane
Coupling agent

| Silane application | Lamination temperature, °C | Flexural strength, MPa Dry | Flexural strength, MPa Wet, 2-hr boil |
|---|---|---|---|
| None | 60 | 482 | 365 |
| None | 105 | 387 | 237 |
| 0.5% Pretreat, room temperature | 60 | 607 | 595 |
| 0.5% Pretreat, | | | |
| 7 min 110°C | 60 | 668 | 641 |
| 7 min 160°C | 105 | 522 | 481 |
| 1% add | 60 | 534 | 428 |
| 1% add | 105 | 639 | 617 |
| 1% add | 105$^a$ | 632 | 571 |

$^a$ Postcure 30 min at 122°C.

The chemical bonding mechanism has been demonstrated in bonding
of silanes to thermosetting resins and bonding of silane to glass or silica,
but several important observations do not support such a simple mechanism.
Although a covalent bridge of bonds may form from resin to mineral through
organofunctional silanes, some covalent oxane bonds (e.g., AlOSi) have
very poor hydrolytic stability, and yet very water-resistant bonds through
silanes to aluminum are possible.[6] Direct vulcanization of rubber against
silane-treated glass produces covalent bonds across the interface, but poor
adhesion generally results unless an added resinous primer is used.[7] Certain
organofunctional silanes are very good coupling agents for inert thermoplastic (e.g., polyolefins) under conditions where no chemical reaction with the
polymer can be demonstrated.[36] Silane-modified tackifying resins are good

**Table 1.9.** Effect of Surface Treatment on Glass Bead–Polyester
Composite Strengths

| Surface treatment | Volume fraction glass beads | Tensile strength, MPa(psi) |
|---|---|---|
| Pure resin | — | 54.7 (7920) |
| Untreated | 0.21 | 40.6 (5890) |
| Methacryloxysilane | 0.22 | 54.9 (7960) |
| Amine and methacryloxysilane | 0.22 | 58.5 (8490) |
| Vinylsilane | 0.22 | 40.5 (5580) |
| Amine and Vinylsilane | 0.22 | 50.9 (7390) |

adhesion promoters for thermoplastic elastomers where no chemical reaction between silane and resin or silane and elastomer is possible.[37] All of these observations will be reconciled with a general mechanism of adhesion in Chapter 5.

### 1.5.2. Wetting and Surface Energy Effects

Sir Isaac Newton, in discussing the attraction between bodies, observed that "parts of all homogeneal hard bodies which fully touch one another, stick together very strongly."[38] In a review of known aspects of surface chemistry and surface energies as related to adhesion, Zisman in 1963 concluded that good wetting of adherend by liquid resin was of prime importance in preparation of composites.[39] Physical adsorption of resin on high-energy surfaces would provide adhesive strength far in excess of cohesive strength of organic resins if complete wetting were obtained.

To obtain complete wetting of a surface, the adhesive must initially be of low viscosity and have a surface tension lower than the critical surface tension $\gamma_c$ of the mineral surface. Although all solid mineral surfaces have very high $\gamma_c$, many hydrophilic minerals in equilibrium with atmospheric moisture are covered with a layer of water. Thus, in a humid atmosphere, poor wetting and spreading of a nonpolar adhesive would occur in contact with the moist surface of a polar adherent. Polar adhesives, on the other hand, may be able either to absorb the water or displace it through surface-chemical reaction. This mechanism can sometimes be enhanced by polar additives to the adhesive.

It was considered of prime importance to study the nature and orientation of silane coupling agents on glass and the effect they have on $\gamma_c$ of the treated surface, in order to provide a surface with a $\gamma_c$ higher than the surface tension of liquid resins. Pure diglycidyl ether of bisphenol-A has a surface tension at 40°C of about 42.5 dyn/cm and is little changed by addition of $m$-phenylenediamine curing agent. Polyester–styrene mixtures would have much lower surface tension. Treated glass, therefore, should preferably have a minimum $\gamma_c$ of about 43 dyn/cm for epoxies and 35 dyn/cm for polyesters.

Six years later, Zisman[40] was able to summarize and evaluate a mass of new data related to the surface chemistry of mineral fibers in reinforced composites. Critical surface tensions of various silane coupling agent films deposited on mineral surfaces were reported by Bascom[41] and Lee.[42] Comparing these values with the $\gamma_c$ of glass in dry and humid atmospheres, it is evident that the wettability of glass can be controlled by treatment with the proper hydrolyzable silanes (Table 1.10). It was somewhat disconcerting, however, to discover that the proven silane coupling agents most commonly

**Table 1.10.** Critical Surface Tensions of Silane Coupling Agent Films[a] (applied to substrate from water)

| Coupling agent structure | Substrate | $\gamma_c$ |
|---|---|---|
| $CF_3(CF_2)_6CH_2O(CH_2)_3Si(OC_2H_5)_3$ | Pyrex glass | 14 |
| $(CF_3)_2CFO(CH_2)Si(OCH_3)_3$ | Stainless steel | 14 |
| | Pyrex glass | 18 |
| | Silica | 17 |
| $CH_3Si(OCH_3)$ | Soda-lime glass | 22.5 |
| $C_2H_5Si(OC_2H_5)_3$ | Silica | 26–33[b] |
| $CH_2{=}CHSi(OC_2H_5)_3$ | Silica | 30 |
| $CH_2{=}CHSi(OCH_3)_3$ | Soda-lime glass | 25 |
| $CH_2{=}C(CH_3)COO(CH_2)_3Si(OCH_3)_3$ | Soda-lime glass | 28 |
| $H_2NCH_2CH_2NH(CH_2)_3Si(OCH_3)_3$ | Soda-lime glass | 33.5 |
| $CH_3C_6H_4Si(OCH_3)_3$ | Soda-lime glass | 34 |
| $H_2N(CH_2)_3Si(OC_2H_5)_3$ | Soda-lime glass | 35 |
| $BrCH_2C_6H_4Si(OCH_3)_3$ | Soda-lime glass | 39.5 |
| $\overset{O}{\frown}{-}CH_2{-}CH{-}CH_2O(CH_2)_3Si(OCH_3)_3$ | Soda-lime glass | 38.5–42.5[b] |
| $C_6H_5Si(OCH_3)_3$ | Soda-lime glass | 40 |
| $Cl(CH_2)_3Si(OCH_3)_3$ | Soda-lime glass | 40.5 |
| | Pyrex glass | 43 |
| | Stainless steel | 44 |
| $HS(CH_2)_3Si(OCH)_3$ | Soda-lime glass | 41 |
| $p\text{-}ClC_6H_4CH_2CH_2Si(OCH_3)_3$ | Pyrex glass | 40–45[b] |
| | Silica | 44 |
| | Stainless steel | 47 |
| $BrC_6H_4Si(OCH_3)_3$ | Soda-lime glass | 43.5 |
| Air at 1% relative humidity | Soda-lime glass | 47 |
| Air at 95% relative humidty | Soda-lime glass | 29 |

[a] From E. P. Plueddemann, Ed., *Composite Materials*, Vol. VI, p. 7, Academic, New York (1974).
[b] Depending on catalyst used.

used in the reinforced plastics industry led to critical surface tensions of less than 35. The best coupling agent for polyester resins (a methacrylate-ester-functional silane) imparted a critical surface tension of only 28. In addition, it was reported that chloropropyl ($\gamma_c = 40.5$) and bromophenyl silanes ($\gamma_c = 43.5$) were completely ineffective as coupling agents in glass-reinforced polyesters.

In fact, no nonreactive silane on glass contributed any improvement in overall performance in polyester composites compared with untreated glass; while, among reactive silanes, the performance paralleled reactivity and not polarity (as described by solubility parameter $\delta$) of the organofunctional groups.[23] For example, comparison of a methacrylate-functional silane and a crotonate-ester silane with similar $\delta$ values of about 9 shows how this difference relates to reactivity. The methacrylate is among the best known silane coupling agents for polyesters, while the crotonate is among the poorest of unsaturated functional silanes in this application. Reactivity of the silane in copolymerization is obviously of much greater significance than polarity or wettability of the treated glass surface (Figure 1.6).

### 1.5.3. Morphology

It has been proposed that silane treatments on reinforcements and fillers in some way modified the morphology of adjacent polymer to improve adhesion.

The probability that orientation may extend beyond the silane finish into the resin phase is suggested by work of Kahn with orientation of liquid crystals on mineral surfaces treated with silane coupling agents.[43] A methylaminopropylsilane on oxide surfaces caused parallel alignment, while an octadecylaminopropylsilane caused perpendicular alignment of liquid crystals in contact with the surface.

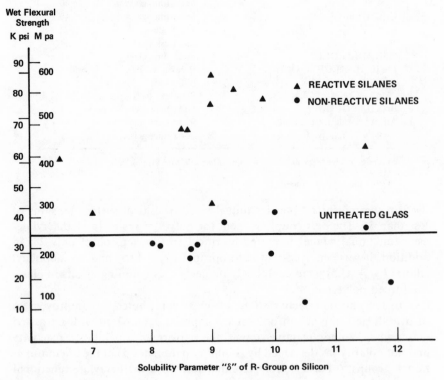

**Figure 1.6.** Polarity of nonreactive and reactive silanes related to their performance in polyester–glass laminates.

**1.5.3.1. Deformable Layer Theories.** There has been considerable confusion, however, over what morphology would be desired in resin adjacent to the treated mineral. A flexible, deformable phase seems desirable to accommodate stresses set up at the interface due to differential thermal shrinkage between resin and filler when the composite is cooled. Broutman and Agarwal[44] estimated that a composite should reach a maximum toughness when the modulus at the interface is about 80 MPa ($10^4$ psi).

The amount of silane in a typical glass finish is not sufficient to provide a low-modulus layer at the interface, but Erickson et al. proposed that the silane-treated surface might allow preferential adsorption of one ingredient in the resin.[45] Unbalanced cure in the interphase region could provide a flexible resin much thicker than the silane layer.

**1.5.3.2. Restrained Layer Theory.** At the other extreme, it has been proposed that the resin in the region of mineral filler should have a modulus somewhere between that of the mineral and the matrix resin.[46] The restrained layer theory suggests that the silane coupling agents function by "tightening up" the polymer structure in the interphase region.

Performance data on reinforced composites suggests that a restrained layer is required at the interface for maximum bonding and resistance to hydrolytic cleavage.

The effect of silane-modified kaolin filler on orientation of high-density polyethylene in the interphase region was reported by Gaehde.[36] Although a vinylsilane grafted to polyethylene while a methacrylate silane homopolymerized at the interface, the methacrylate silane was more effective in increasing the crystallinity of polyethylene at the interface. In addition, it is believed that methacrylate silanes and aminoalkyl silanes form interpenetrating polymer networks with polyolefins at the interface.[8] The vinyl silane was the least effective coupling agent of the three, even though it was the only one that reacted chemically with polyethylene (Table 1.11). Mechanical properties of kaolin-filled polyethylene thus seemed to be more dependent on polymer structure at the interface than on chemical grafting.

**Table 1.11.** Silane-Treated Kaolin in High-Density Polyethylene

| Silane on kaolin | Mode of action | Strengths of composites (20 vol% clay), MPA | |
| --- | --- | --- | --- |
| | | Tensile strength | Flexural strength |
| None | — | 190 | 369 |
| Vinyl silane | Graft | 200 | 379 |
| Methacrylate silane | Orient and IPN | 238 | 483 |
| Aminopropyl silane | IPN | 221 | 407 |

Fillers are known to have varying degrees of catalytic effect on thermo-setting resins that generally inhibit their cure. Exotherm measurements[32] and direct observations of functional groups with FT-IR indicate that glass, silica, and many other mineral surfaces inhibit the cure of polyester and epoxy resins.[31]

In both polyesters and epoxies, it was observed that silane treatments of fillers often overcome cure inhibition as measured by cure exotherms. Silanes that allowed maximum exotherms, that is, were most effective in overcoming surface inhibition, were generally the most effective coupling agents as indicated by mechanical properties and chemical resistance of composites. A "restrained layer," rather than a "deformable layer," must therefore be the best morphology in the interphase region for optimum properties of composites.

**1.5.3.3. Tough Composites from Rigid Resins.** It may be concluded that maximum toughness of composites is obtained with a deformable layer while maximum chemical resistance is obtained with a restrained layer at the interface. Tough, water-resistant composites may be prepared from rigid resins by bonding a thin elastomeric interlayer to the mineral surface through a cross-linkable silane. Elastomeric interlayers are selected that bond to the matrix resin by interdiffusion or some related mechanism.[47] Thus, an SBS block copolymer interlayer bonds well to polyolefins, while an elastomeric urethane bonds well to polycarbonates, polyterephthalates, and polyamides.

The complete composite of silane primer on mineral substrate (e.g., glass, steel, aluminum), elastomeric interlayer, and rigid matrix resin provides a structure with improved toughness while retaining good adhesion and water resistance (cf. Section 6.6).

### 1.5.4. Acid–Base Reactions

It is common practice to modify polymers with carboxyl groups for improved adhesion to metals.[48] Bolger pointed out that various metal oxide surfaces have different isoelectric points in water and therefore might be considered as more or less acid or basic. It might be expected, therefore, that modification of an adhesive with acid groups should improve adhesion to basic surfaces.[49] Acid–base reactivity between filler and matrix could be further modified by adding suitable acid or basic solvents to the system.[50] It is also known that acid–base reactions may be important in aligning silane coupling agents on a filler surface.[32]

Acid–base reactions between dissimilar polymers may be very impor-tant in improving interpolymer adhesion. Modification of inorganic sub-strates may also improve acid–base reactions between polymers and reinforcements.[51] Acid-functional polymers develop good dry adhesion to glass and metals, but the bonds have limited resistance to moisture.

To determine whether acid–base reactions were important in developing water-resistant bonds between polymers and metal oxides, a series of functional modified acrylic varnishes were evaluated[52] for wet-adhesion to mineral surfaces ranging in surface isoelectric points from 2 to 12. Four mol percent of functional methacrylate monomers were copolymerized with methylmethacrylate in solution. The resulting varnishes were coated on the various surfaces, dried for 30 min at 100°C, and immersed in water at room temperature. Adhesion was observed by attempting to pry films from the surface with a razor blade. Samples were rated qualitatively from 0 to 10, according to how long they withstood water before the films could be loosened. A rating of 0 indicated poor initial adhesion, while a rating of 10 indicated no loss in adhesion after 7 days in water (Table 1.12). Copolymers of 4 mol% functional methacrylates with styrene showed similar patterns of adhesion. These data indicate that any polar functional group improved adhesion. There was little evidence of improved water resistance through acid–base reactions across the interface. At this low level of modification, the functional silanes were in a class far above any other organofunctional modifiers. It must be concluded that, although acid–base reactions may be important in filler dispersion, in orientation of silanes on surfaces, and in improvement in dry adhesion, the major factor in water-resistant bonds through silanes is something other than acid–base reactions across the interface.

A modified acid–base mechanism was described in which acid-functional silanes on the reinforcement bond to acid groups in the polymer in the presence of metal ions through ionomer bonds. Ionomer are fluid under

**Table 1.12.** Wet-Adhesion of Methylmethacrylate Copolymers (4 mol% functional modification)

| Function in polymer | Adhesion to surface (IEPS) | | | | | | | |
|---|---|---|---|---|---|---|---|---|
|  | Si (2) | Sn (4) | Glass (5) | Ti (6) | Cr (7) | Al (9) | Steel (10) | Mg (12) |
| None | 1 | 0 | 1 | 1 | 2 | 2 | 2 | 2 |
| —OH | 5 | 3 | 3 | 3 | 3 | 3 | 5 | 3 |
| —COOH | 2 | 3 | 2 | 5 | 2 | 3 | 5 | 2 |
| —CONH$_2$ | 3 | 2 | 2 | 3 | 3 | 8 | 3 | 10 |
| —CH$_2$CHCH$_2$O | 8 | 1 | 2 | 3 | 2 | 10 | 3 | 3 |
| —NMe$_2$ | 3 | 2 | 2 | 3 | 2 | 3 | 5 | 3 |
| —Si(OAc)$_3$ | 10 | 10 | 10 | 10 | 10 | 10 | 10 | 3 |

Rating: 0 = no adhesion; 10 = Good adhesion after 7 days in water.

conditions of high temperature and high shear, but solidify at room temperature to tough water-resistant structures. Structures in which 30–40% of carboxy groups were neutralized by $Na^+$ or $Zn^{2+}$ (or other cations) produced very water-resistant bonds between glass and polyethylene, polypropylene, nylon and other thermoplastic polymers.[11]

### 1.5.5. Other Theories

Many other functions may be served by coupling agents at a resin-mineral interface. The coupling agent may provide lubrication to protect the mineral against abrasion during fabrication.[53] Such protection is especially important with glass fibers. The coupling agents may protect the mineral surface against stress corrosion by water.[54] It is difficult to demonstrate whether stress corrosion by water causes mechanical failure at the interface, or whether mechanical failure at the interface allows the entrance of water necessary to initiate stress corrosion. In some cases, it seems that a silane finish actually heals flaws on the glass surface.[55]

It appears that all the theories of adhesion describe factors that are involved in bonding through silane coupling agents. An overall picture of adhesion through silane coupling agents is developed in Chapter 5.

## References

1. Plato. *Timaeus*, English translation by R. D. Archer Hinds, McMillan, London (1888).
2. P. C. Yates and J. W. Trebilock. SPI, 16th Ann. Tech. Conf. Reinf. Plast. 8-B (1961).
3. J. Bjorksten and L. L. Yaeger. *Mod. Plast.* **29**, 124 (1952).
4. E. P. Plueddemann, H. A. Clark, L. E. Nelson, and K. R. Hoffmann. *Mod. Plast.* **39**, 136 (1962).
5. H. Ishida and J. L. Koenig. *J. Polym. Sci., Polym. Phys. Ed.* **17**(10), 1807 (1979).
6. E. P. Plueddemann, *J. Adhesion Sci. Tech.* **2**(3), 179 (1988).
7. E. P. Plueddemann and W. T. Collins. *Adhesion Science and Technology*, L. H. Lee, Ed., Part A, pp. 329–353, Plenum, New York (1975).
8. E. P. Plueddemann and G. L. Stark. SPI, 35th Ann. Tech. Conf. Reinf. 20-B (1980).
9. B. T. Vanderbilt and R. E. Clayton. *Ind. Eng. Chem., Prod. Res. Dev.* **4**(1), 16 (1965).
10. D. J. Vaughan (Technical Director, Clark Schwebel Fiberglass Corporation). Private communication (1971).
11. E. P. Plueddemann. *J. Adhesion Sci. Tech.* **3**(2), 131–139 (1989).
12. J. F. Mandell, D. H. Grande, T. H. Tsiang, and F. J. McGarry. In *Composite Materials: Testing and Design (Seventh Conference)*, ASTM STP 893, p. 87 (1986).
13. D. L. Caldwell and D. A. Jarvie. 33rd Int. SAMPE Symposium, Anaheim, CA, p. 1268 (1988).
14. W. A. Fraser, F. H. Ancker, and A. T. Dibenedetto. SPI, 30th Ann. Tech. Conf. Reinf. Plast. 22-A (1975).
15. L. T. Drzal, M. J. Rich, J. D. Camping, and W. J. Park. SPI, 35th Ann. Tech. Conf. Reinf. Plast. 20-C (1980).

16. P. Yeung and L. L. Broutman. SPI, 32nd Ann. Tech. Conf. Reinf. Plast. 9-B (1977).
17. F. Hoersch. *Kunststoffe* **55**, 909 (1965).
18. R. F. Regester. SPI, 22nd Ann. Tech. Conf. Reinf. Plast. 16-D (1967).
19. R. R. Levell. SPI, 22nd Ann. Tech. Conf. Reinf. Plast. 16-D (1967).
20. P. C. Yates and J. W. Trebilcock. SPI, 16th Ann. Tech. Conf. Reinf. Plast. 8-B (1961).
21. C. Q. Yang and Q. L. Zhou. *Appl. Surf. Sci.* **23**, 213 (1985).
22. P. Erickson and I. Silvers, U.S. Patent 2,776,910 (1956).
23. E. P. Plueddemann. *J. Paint Technol.* **40** (516), 1 (1968).
24. K. M. Foley and M. Vigo. U.S. Patent 3,944,707 (to Owens Corning Fiberglass). (1976).
25. M. E. Schrader and I. Lerner. U.S. Patent 3,420,726 (to U.S. Navy) (1969).
26. M. M. Feing, B. K. Patnaik, and F. K. Y. Chu. U.S. Patent 4,073,766 (to Dart Industries) (1978).
27. A. A. Pike, F. P. Lamm, and J. P. Pinto. Proceedings 5th Int. Joint Military/Govt. Industry Symp. on Adhes. Bonding, Picatinny Arsenal, p. 177 (1987).
28. L. B. Cohen. SPI, 41st Ann. Tech. Conf. Reinf. Plast. 26-A (1986).
29. S. J. Monte and G. Sugarman. In *Adhesion Aspects of Polymeric Coatings*, K. L. Mittal, Ed, p. 421, Plenum, New York (1983).
30. D. E. Cope. SPI, 33rd Ann. Tech. Conf. Reinf. Plast. 5-A (1978).
31. H. Ishida and J. L. Koenig. *J. Polym. Sci.* **17**, 615 (1979).
32. E. P. Plueddemann and G. L. Stark. *Mod. Plast.* **51**(3), 74 (1974).
33. H. A. Clark and E. P. Plueddemann. *Mod. Plast.* **40**(6), 133 (1963).
34. R. L. Kaas and J. L. Kardos. *Soc. Plast. Eng., Tech. Pap.* **22**, 22 (1976).
35. H. Ishida and J. L. Koenig. *J. Colloid Interf. Sci.* **64**(3), 565 (1978).
36. J. Gaehde. *Plaste Kauteschuk* **22**(8), 626 (1975).
37. E. P. Plueddemann. Applied Polymer Symposium, No. 19, pp. 25–90, Wiley, New York (1972).
38. I. Newton. *Optics*, 2nd ed, London, p. 350 (1978).
39. W. A. Zisman. *Ind. Eng. Chem.* **35**(10), 19 (1963).
40. W. A. Zisman. *Ind. Eng. Chem., Prod. Res. Devel.* **8**(2), 98 (1969).
41. W. D. Bascom. *J. Colloid Interface Sci.* **27**, 789 (1968).
42. L. Hn. Lee. *J. Colloid Interface Sci.* **27**, 751 (1968).
43. F. J. Kahn. *Appl. Phys. Let.* **22**(8), 386 (1973).
44. L. J. Broutman and B. D. Agarwal. SPI, 28th Ann. Tech. Conf. Reinf. Plast. 5-B (1973).
45. P. W. Erickson, A. A. Volpe, and E. R. Cooper. SPI, 19th Ann. Tech. Conf. Reinf. Plast. 24-A (1974).
46. C. A. Kumins and J. Roteman. *J. Polymer. Sci., Part A* **1**, 527 (1963).
47. E. P. Plueddemann. SPI, 29th Ann. Tech. Conf. Reinf. Plast. 24-A (1974).
48. F. M. Rosenblum. *Adhesives Age* **22**(4), 20 (1979).
49. J. C. Bolger and A. S. Michaels. *Interface Conversion for Polymer Coatings*, P. Weiss and G. D. Cheeves, Eds., pp. 2–60, Elsevier, New York (1969).
50. M. J. Marmo, M. A. Mostafa, H. Jinnai, F. M. Fowkes, and J. A. Manson. *Ind. Eng. Chem., Prod. Res. Dev.* **15**(3), 206 (1976).
51. F. M. Fowkes, *J. Adhesion Sci. Tech.* **1**(1), 7 (1987).
52. E. P. Plueddemann and G. L. Stark. SPI, 28th Ann. Tech. Conf. Reinf. Plast. 21-E (1973).
53. J. O. Outwater. SPI, 11th Ann. Tech. Conf. Reinf. Plast. 9-B (1956).
54. R. L. Patrick, W. G. Gahman, and L. Dunbar. *J. Adhesion* **3**(2), 165 (1971).
55. D. J. Vaughan and J. W. Sanders, Jr. SPI, 29th Ann. Tech. Conf. Reinf. Plast. 13-A (1974).

# 2 | Chemistry of Silane Coupling Agents

## 2.1. Introduction

The "coupling" mechanism of organofunctional silanes depends on a stable link between the organofunctional group (Y) and hydrolyzable groups (X) in compounds of the structure $X_3SiRY$. The organofunctional groups are chosen for reactivity or compatibility with the polymer, while the hydrolyzable groups are merely intermediates in formation of silanol groups for bonding to mineral surfaces.

Vinylsilanes were the first commercial silane coupling agents used with reinforced unsaturated polyesters. It was demonstrated in fiberglass-reinforced polyester composites that $ViSiX_3$ compounds with various hydrolyzable X groups were essentially equivalent when applied to glass. Best wet-strength retention was obtained with silanes having three hydrolyzable groups on silicon[1] (Table 2.1).

Organofunctional groups attached to silicon through alkoxy or acyloxy linkages generally do not have sufficient hydrolytic stability to provide environmentally stable bonds between resin and reinforcement. Thus, allyoxysilanes or silanes with glycidoxy or methacryloxy substituents on silicon are not useful as coupling agents.

An exception may be found in functional alkoxysilanes having a negative substituent on the second carbon of the alkoxy group. Reaction products of glycidyl methacrylate with chlorosilanes contain a $SiOCH_2CHClCH_2Y$ (Y = methacryloxy) linkage that contributes very creditable water resistance to reinforced polyesters.[2] Comparable compounds of germanium, tin, titanium, and zirconium were much less effective as coupling agents.[3]

**Table 2.1.** Vinylsilane Finishes in
Polyester Laminates

| | Flexural strength, MPa | |
| Silane in finish | Dry | 2-hr water boil |
|---|---|---|
| None (control) | 387 | 240 |
| ViSi(O)ONa | 400 | 410 |
| ViSi(O)ONMe$_4$ | 390 | 410 |
| ViSi(OMe)$_3$ | 464 | 416 |
| ViSi(OEt)$_3$ | 413 | 405 |
| ViSi(OCH$_2$CH$_2$OMe)$_3$ | 472 | 441 |
| ViSi(Ac)$_3$ | 441 | 455 |
| ViSiCl$_3$ | 496 | 408 |
| ViMeSiCl$_2$ | 285 | 324 |
| ViMe$_2$SiCl | 419 | 351 |

## 2.2. Formation of the Si—C Bond

Most of the standard methods for forming Si—C bonds may be used in preparing intermediates for silane coupling agent synthesis. Standard texts such as those by Eaborn[4] or Noll[5] may be consulted. The addition of silicon hydrides to substituted olefins and acetylenes is the most important laboratory and industrial process:

$$X_3SiH + CH_2{=}CH{-}R{-}Y \rightarrow X_3SiCH_2CH_2RY$$

$$X_3SiH + CH{\equiv}CH \rightarrow X_3SiCH{=}CH_2$$

The additions may take place in the liquid or gas phase, by merely heating the reagents together. The reactions are better done in the presence of catalysts such as peroxides, tertiary bases, or platinum salts.[6] Catalyzed addition reactions are especially important in attaching to silicon organic radicals that contain reactive functional groups, because such groups usually cannot be introduced by organometallic procedures or by direct addition at high temperatures.

At high temperatures, trichlorosilane reacts with benzene or toluene to give trichlorosilyl substitution products:

$$HSiCl_3 + C_6H_6 \xrightarrow{350°} PhSiCl_3$$

$$HSiCl_3 + CH_3C_6H_5 \xrightarrow{350°} CH_3C_6H_4SiCl_3$$

Because $HSiCl_3$ is highly volatile and corrosive, and $HSi(OMe)_3$ is unstable and highly toxic, it is often inconvenient to synthesize organofunctional silanes in the laboratory from such basic ingredients. Fortunately, there are available a goodly number of basic commercial organofunctional silanes that may be modified by standard organic reactions to prepare almost any conceivable organofunctional silane. Some representative organofunctional silanes available in commercial quantities are shown in Table 2.2. Many additional compounds are available as laboratory chemicals from companies (e.g., Petrarch Systems, PCR, Inc.) specializing in research chemicals.

Both ends of the silane molecule $X_3SiRY$ may undergo chemical reactions, either separately or simultaneously. With proper control of conditions, the X groups can be replaced without altering the Y group, or the Y group may be modified while retaining the X group. If the Y group is modified in an aqueous environment, the X groups are simultaneously hydrolyzed. Chemical reactions of the Y group may precede application to a surface or may take place at a surface after silylation.

## 2.3. Reactions of Hydrolyzable Groups on Silicon

### 2.3.1. Alcoholysis

Alkoxysilanes are commonly prepared by alkoxylation of chlorosilanes. This reaction proceeds readily without catalyst, but it requires efficient removal of hydrogen chloride. Evolution and recovery of anhydrous HCl is preferred in commercial practice. Laboratory preparations may include HCl acceptors like tertiary amines or sodium alkoxides. A convenient method of completing the alkoxylation reaction is by warming a chlorosilane with the appropriate orthoformate ester in the presence of alcohol.[7] One mole of orthoformate is consumed for each equivalent of chloride removed.

$$\equiv SiCl + HC(OR)_3 \xrightarrow{ROH} \equiv SiOR + RCl\uparrow + HCOOR$$

Alkoxy exchange of $RSi(OR)_3$ requires a catalyst to be efficient at moderate temperatures. The catalyst may be a mineral acid, a Lewis acid (e.g., alkyltitanate), or a strong base. Reaction conditions must be adapted to the organofunctional group on silicon. Thus, glycidoxy propyltrimethoxysilane may be alcoholized in the presence of a base catalyst, while acid catalysts must be avoided if the epoxy group is to be retained.[8] The amine group of aminopropyltrialkoxysilanes is not sufficiently basic to catalyze rapid alcoholysis of these silanes. A trace of sodium alkoxide is a very effective catalyst for this reaction.

$$H_2N(CH_2)_3Si(OCH_3)_3 + 3ROH \xrightarrow{NaOR} 3CH_3OH + H_2N(CH_2)_3Si(OR)_3$$

**Table 2.2.** Representative Commercial Silanes

| Organofunctional group | Chemical structure | Abbreviations |
|---|---|---|
| A. Vinyl | $CH_2=CHSi(OCH_3)_3$ | VS |
| B. Chloropropyl | $ClCH_2CH_2CH_2Si(OCH_3)_3$ | CPS |
| C. Epoxy | $CH_2CHCH_2OCH_2CH_2CH_2Si(OCH_3)_3$ <br> (epoxy ring on $CH_2CHCH_2$) | GPS |
| D. Methacrylate | $CH_2=\overset{\underset{\displaystyle CH_3}{\mid}}{C}-COOCH_2CH_2CH_2Si(OCH_3)_3$ | MPS |
| E. Primary amine | $H_2NCH_2CH_2CH_2Si(OC_2H_5)_3$ | APS |
| F. Diamine | $H_2NCH_2CH_2NHCH_2CH_2CH_2Si(OCH_3)_3$ | AEAPS |
| G. Mercapto | $HSCH_2CH_2CH_2Si(OCH_3)_3$ | MGPS |
| H. Cationic styryl | $CH_2=CHC_6H_4CH_2NHCH_2CH_2NH(CH_2)_3Si(OCH_3)_3 \cdot HCl$ | CSS |

Silicone-modified alkyds or acrylics are prepared by cooking alkoxy-functional silicones with hydroxyl-containing resins. Alkyl-titanate catalysts are recommended for alcoholysis to form copolymers of silicones and the organic resins. Silanol functional silicones are cooked directly with hydroxy-functional resins to form copolymers without a catalyst.

### 2.3.2. Acylation

Chlorosilanes react with sodium acetate in anhydrous solvents to form acetoxysilanes:

$$RSiCl_3 + 3NaAc \rightarrow RSi(OAc)_3 + 3NaCl$$

Formation of precipitated salt may be avoided by heating the chlorosilane with acetic anhydride and removing volatile acetylchloride:

$$RSiCl_3 + 3Ac_2O \rightarrow RSi(OAc)_3 + 3AcCl\uparrow$$

Acetoxysilanes are more reactive with water than alkoxysilanes and have some advantage in ease of forming aqueous solutions. The major applications of acetoxysilanes, however, are as cross-linkers in moisture-reactive room-temperature-vulcanizable (RTV) silicone elastomers.

### 2.3.3. Hydrolysis and Condensation

Trialkoxysilanes, $RSi(OR^1)_3$, hydrolyze stepwise in water to give the corresponding silanols, which ultimately condense to siloxanes. Both reaction rates are strongly dependent on pH, but under optimum conditions

$$RSi(OMe)_3 \xrightarrow{H_2O} RSi(OH)_3 \rightarrow HO-\underset{\underset{OH}{|}}{\overset{\overset{R}{|}}{Si}}-\left(O-\underset{\underset{OH}{|}}{\overset{\overset{R}{|}}{Si}}-\right)_n O-\underset{\underset{OH}{|}}{\overset{\overset{R}{|}}{Si}}-OH$$

the hydrolysis is relatively fast (minutes), while the condensation reaction is much slower (several hours). Higher alkoxysilanes hydrolyze very slowly in water because they are strongly hydrophobic, but even in homogeneous solution in water-miscible solvents, they hydrolyze more slowly than the lower alkoxysilanes.[9]

Alkoxysilane groups of silane coupling agents may react directly with silanol groups of a siliceous surface,[10] although a catalyst such as an alkyl titanate or an amine is often recommended to accelerate the condensation.[11]

Silane coupling agents may also be prehydrolyzed and applied to siliceous surfaces from aqueous solutions. Under these conditions, silanol groups of the coupling agent condense with hydroxyl groups of the mineral surface during drying operations.

The importance of reactive groups on silicon in coupling agents (alkoxy or silanol) is demonstrated by the observation that methacrylate-functional siloxanes applied to glass fibers from toluene solution were not effective coupling agents in polyester laminates, while methacrylate-functional silanols or methyoxysilanes are mong the most effective coupling agents for polyester.[1]

The nature of aqueous solutions of silane coupling agents is discussed in detail in Chapter 3.

## 2.4. Organofunctional Groups on Silicon

### 2.4.1. Halogen

**2.4.1.1. Preparation.** Chloropropyltrimethyoxysilane is available commercially and may be used as an intermediate in synthesis. Other halofunctional silicon compounds may be available from laboratory chemical suppliers or may be synthesized from commercial materials.

Methyltrichlorosilane may be chlorinated to chloromethyltrichlorosilane:

$$CH_3SiCl_3 + Cl_2 \xrightarrow{\text{light}} ClCH_2SiCl_3 + HCl$$

To avoid the formation of polychloromethylsilanes, a continuous recycling procedure is used in which the more volatile material is chlorinated in the vapor phase.[12]

The chloromethyl group is cleaved from silicon by strong alkali, and, in general products derived from chloromethylsilanes, shows no advantage over those derived from chloropropylsilanes:

$$\equiv SiCH_2Cl \xrightarrow[C_2H_5OH]{\text{KOH}} \equiv SiOC_2H_5 + CH_3Cl$$

It is generally best to avoid 2-chloroethylsilicon compounds because they cleave with extreme ease; the chlorine may be titrated at room temperature with 0.1 $N$ base[13]:

$$\equiv SiCH_2CH_2Cl \xrightarrow{NaOH} \equiv SiOH + CH_2{=}CH_2 + NaCl$$

A 3-bromopropylsilane is made by the addition of trichlorosilane to allylbromide. Its major utility is in analytical studies of silane coupling mechanisms because of the clear signal given by bromine in X-ray fluorescence:

$$HSiCl_3 + CH_2{=}CHCH_2Br \rightarrow Cl_3SiCH_2CH_2CH_2Br$$

Iodoalkylsilanes are best prepared by replacement reactions of chloroalkylsilanes with NaI:

$$(MeO)_3SiCH_2CH_2CH_2Cl + NaI \xrightarrow{acetone}$$

$$(MeO)_3SiCH_2CH_2CH_2I + NaCl$$

A more active chlorofunctional silane is obtained by adding trichlorosilane to the double bond of vinylbenzyl chloride:

$$HSiCl_3 + CH_2{=}CHC_6H_4CH_2Cl \xrightarrow{Pt} Cl_3SiCH_2CH_2C_6H_4CH_2Cl$$

Coupling agents derived from this product are reported to perform better in high-temperature resin composites than conventional aliphatic silanes.

Aromatic silanes may be halogenated in the ring to form haloarylsilanes. These products have been incorporated into silicone fluid polymers to improve lubricity, but they have not found application as coupling agents:

$$Cl_3SiC_6H_5 + Cl_2 \xrightarrow{Fe} Cl_3SiC_6H_4Cl + HCl$$

**2.4.1.2. Reactions.** Haloalkylsilanes may be used as coupling agents in composites where reaction with the resin is possible at fabrication temperatures. Thus, chloropropylsilanes are effective coupling agents with polystyrene (in presence of a trace of $FeCl_3$)[15] or with high-temperature-cured epoxies, but not with unsaturated polyesters.

Haloalkylsilanes find greatest utility as intermediates in the preparation of other organofunctional silanes. They react readily with ammonia or amines to form amino-organofunctional silanes.[16] With tertiary amines they form quaternary ammonium compounds that are useful for surface-bonded antimicrobial activity on textiles or inorganic surfaces.[17] With sodium alkoxides they form ethers[18]:

$$(MeO)_3SiCH_2CH_2CH_2Cl + 2H_2NR \rightarrow$$

$$(MeO)_3SiCH_2CH_2CH_2CH_2NHR + RNH_2-HCl$$

$$(MeO)_3SiCH_2CH_2C_6H_4CH_2Cl + NaOMe \rightarrow$$

$$(MeO)_3SiCH_2CH_2C_6H_4CH_2OCH_3 + NaCl$$

Alkali or tertiary amine salts of carboxylic acids react with haloalkylsilanes to form esters[14]:

$$(MeO)_3SiCH_2CH_2CH_2Cl + CH_2=C(CH_3)COOH \xrightarrow{R_3N}$$

$$CH_2=C(CH_3)COOCH_2CH_2CH_2Si(OMe)_3$$

The esters themselves may be valuable coupling agents or they may be alcoholized to form hydroxyalkylsilanes:

$$(MeO)_3SiCH_2CH_2CH_2OAc + MeOH \xrightarrow{HCl}$$

$$(MeO)_3SiCH_2CH_2CH_2OH + MeOAc$$

Silylalkyl esters of phosphorus are prepared by heating chloroalkyl-silanes with methyl esters of phosphorus in the presence of tertiary amines.[20] In this reaction chloropropyltrimethoxysilane is considerably more reactive than a primary organic chloride of comparable molecular weight:

$$(MeO)_3SiCH_2CH_2CH_2Cl + MeP(O)(OMe)_2 \xrightarrow{R_3N}$$

$$(MeO)_3SiCH_2CH_2CH_2OP(O)(CH_3)OMe + MeCl\uparrow$$

Saponification of the products provides a phosphonate-functional siliconate that has interesting properties as a stabilizer for aqueous silicates (Chapter 3):

$$(MeO)_3SiCH_2CH_2CH_2OP(O)(CH_3)OMe + NaOH \rightarrow$$

$$O_{1.5}SiCH_2CH_2CH_2OP(O)(CH_3)ONa$$

A similar reaction of chloropropyltrimethoxysilane with dimethyl-maleate is accompanied by rearrangement to produce trimethoxysilylpropyl fumarates.[21] Because the methyl maleate esters are more reactive than methyl fumarates, there is a tendency for the reaction to form the monoreaction product:

$$(MeO)_3SiCH_2CH_2CH_2Cl + MeOOCCH{=}CHCOOMe(cis) \xrightarrow{R_3N}$$

$$(MeO)_3SiCH_2CH_2CH_2OOCCH{=}CHCOOMe(trans)$$

Mercaptoalkylsilanes may be prepared from chloroalkylsilanes and ammonium salts of hydrogen sulfide.[22] Simplified recovery of product (by separation of liquid layers) was claimed when the ethylenediamine salt of $H_2S$ was used as reactant[23]:

$$(MeO)_3SiCH_2CH_2CH_2Cl + H_2NCH_2CH_2NH_2{-}H_2S \rightarrow$$

$$(MeO)_3SiCH_2CH_2CH_2SH + H_2NCH_2CH_2NH_2{-}HCl{\downarrow}$$

Chloroalkylsilanes react readily with thiourea to form isothiouronium salts which are cleaved by ammonia to form mercaptoalkylsilanes free of dialkylsulfide byproducts.[24]

$$(MeO)_3SiCH_2CH_2CH_2Cl + H_2NCSNH_2 \rightarrow$$

$$(MeO)_3SiCH_2CH_2CH_2SC(NH)NH_2{-}HCl(I)$$

$$(I) + NH_3 \rightarrow (MeO)_3SiCH_2CH_2CH_2SH + H_2NCONH_2 + NH_4Cl$$

Chloroalkylsilanes have been used in malonic ester and related syntheses[25]:

$$(MeO)_3SiCH_2Cl + NaHC(COOC_2H_5)_2 \rightarrow$$

$$(MeO)_3SiCH_2CH(COOC_2H_5)_2 + NaCl$$

Chloroalkylsilanes also undergo simple replacement reactions with NaCN, KCNS, KCNO, etc.[26]:

$$(MeO)_3SiCH_2CH_2CH_2Cl + KNCO \xrightarrow{DMF}$$

$$(MeO)_3SiCH_2CH_2CH_2NCO + KCl$$

Azides prepared by direct reaction of chloroalkyl silanes and sodium azide are too unstable to be of practical interest as commercial products. Certain sulfonyl azides have a half-life of about 1 hr at 150°C and are recommended as coupling agents for polyolefins.[27] Azidosilane-treated fillers are stable at room temperature, but react with molten thermoplastics through insertion of intermediate nitrenes into any available C—H bond:

$$Si—R—SO_2N_3 \xrightarrow{heat} Si—R—SO_2\ddot{N}: + N_2$$

$$Si—R—SO_2\ddot{N}: + H—CR_3 \rightarrow Si—R—SO_2NH—CR_3$$

Chloropropylsilanes undergo an unusual cleavage reaction at high temperatures with formation of cyclopropane.[28]

$$\equiv SiCH_2CH_2CH_2Cl \xrightarrow{\Delta} \equiv SiCl + cyclo\text{-}(CH_2)_3$$

### 2.4.2.  Carbon=Carbon Double Bonds

**2.4.2.1.  Preparation.** Vinyl trichlorosilane is prepared by the monoaddition of trichlorosilane to acetylene. An excess of acetylene is maintained to minimize bis addition reactions. Trichlorosilane also reacts with allylchloride or vinyl chloride at high temperatures to give unsaturated silanes:

$$HSiCl_3 + HC{\equiv}CH \rightarrow Cl_3SiCH{=}CH_2$$

$$HSiCl_3 + CH_2{=}CHCl \xrightarrow{550-650°} Cl_3SiCH{=}CH_2 + HCl$$

Conjugated dienes undergo 1,4-addition with formation of 3-alkenyl-silanes.

$$HSiCl_3 + CH_2{=}CHCH{=}CH_2 \xrightarrow{Pt/C} Cl_3SiCH_2CH{=}CHCH_3$$

Addition of trichlorosilane to nonconjugated terminal diolefins, followed by addition of diphenylphosphine to the unsaturated product, provided phosphine-functional silanes for binding of metal catalysts to mineral surfaces[31]:

$$HSiCl_3 + CH_2{=}CH(CH_2)_4CH{=}CH_2 \rightarrow Cl_3Si(CH_2)_6CH{=}CH_2 (I)$$

$$(I) + Ph_2PH \rightarrow Cl_3Si(CH_2)_8PPh_2$$

Monoaddition of silanes to divinylbenzene produces styrylethylsilanes showing good performance as coupling agents for unsaturated polyesters.[32]

Silanes add preferentially to the allyl group in allyl esters of unsaturated acids[33] like acrylic, methacrylic, maleic, fumaric, itaconate, etc. The most important of these is the silane derived from allyl methacrylate[34]:

$$HSi(MeO)_3 + CH_2{=}C(CH_3)COOCH_2CH_2{=}CH_2 \rightarrow$$

$$CH_2{=}C(CH_3)COO(CH_2)_3Si(OMe)_3$$

Methallyltrichlorosilane is recovered in good yield in a high-temperature reaction of trichlorosilane with commercial diisobutylene[35]:

$$HSiCl_3 + (CH_3)_3CCH_2C(CH_3){=}CH_2 \xrightarrow{500°}$$

$$CH_2{=}C(CH_3)CH_2SiCl_3 + (CH_3)_3CH$$

Unsaturated amines react with chloroalkylsilanes to form unsaturated cationic-functional silanes[36]:

$$(MeO)_3SiCH_2CH_2CH_2Cl + CH_2{=}C(CH_3)COOCH_2CH_2NMe_2 \rightarrow$$

$$CH_2{=}C(CH_3)COOCH_2CH_2N^{\oplus}(Me)_2(CH_2)_3Si(OMe)_3Cl^-$$

In a similar reaction, vinylbenzyl chloride with amino-organofunctional silanes forms cationic styrylmethyl-functional silanes[37]:

$$(MeO)_3SiCH_2CH_2CH_2NH_2 + ClCH_2C_6H_4CH{=}CH_2 \rightarrow$$

$$(MeO)_3Si(CH_2)_3NHCH_2C_6H_4CH{=}CH_2 \cdot HCl$$

**2.4.2.2. Uses.** Unsaturated silanes are used primarily as coupling agents, but may also be used as chemical intermediates. Vinylfunctional silanes have generally been displaced by methacrylate-functional silanes as commercial coupling agents for unsaturated polyesters, but find extensive use in filled polyethylene where they impart improved electrical properties to cable coverings.[38] Styrene-functional cationic silanes derived from vinyl-benzyl chloride are unique in that they are effective coupling agents with virtually all thermoset resins and thermoplastic resins.[39] With nonreactive thermoplastics, this silane is believed to bond across the interface through an interpenetrating polymer network.[40]

Vinyltrimethoxysilane may be grafted to polyethylene at 140° in the presence of free-radical initiators. The grafted thermoplastic polymer may be cross-linked by subsequent exposure to water in the presence of a condensation catalyst.

**2.4.2.3. Reactions.** Mercaptoalkyltrialkoxysilanes are prepared by reacting hydrogen sulfide with unsaturated trialkoxysilanes in the presence of ultraviolet radiation promoted by trimethylophosphite[42]:

$$(MeO)_3SiCH{=}CH_2 + H_2S \xrightarrow{\text{(MeO)}_3\text{P}} (MeO)_3SiCH_2CH_2SH$$

Addition of mercaptans to vinylsilanes requires free-radical initiators at moderately elevated temperatures. Thus, thioglycolic acid with VS (Table 2.2) in the presence of 0.5% azo-bis isobutyronitrile in 30 min at 100°C gives near to a quantitative yield of the carboxy functional adduct:

$$(MeO)_3SiCH{=}CH_2 + HSCH_2COOH \xrightarrow{\text{azo initiator}}$$

$$(MeO)_3SiCH_2CH_2SCH_2COOH$$

Mercaptans add very readily to the activated double bonds of the methacrylate-functional silane MPS in the presence of catalytic amounts of base. Thus, a mixture of a polysulfide polymer (Thiokol® LP-10) with a slight excess of MPS in the presence of a trace of sodium methoxide gave rapid addition at room temperature, as evidenced by replacement of the typical polysulfide odor with a mild ester odor. The modified polysulfide could be cured to a RTV elastomer by methoxysilane condensation with atmospheric moisture.[43]

Vinylsilanes copolymerize readily with vinyl esters, vinyl chloride, maleic anhydride, allyl esters, maleates, fumarates, and vinyl pyrollidone, but poorly with methyl-methacrylate and styrene. Vinylsilane copolymers are generally of rather low molecular weight, indicating some kind of degradative chain transfer during polymerization. Copolymers of vinyl-trimethoxysilane with maleic anhydride may be used as acid-functional silanes or converted to acid esters or amic acids by reaction with alcohols or amines.[44]

Methacrylate-functional silanes copolymerize readily with styrene and acrylic monomers. Liquid acrylic polymers were prepared by copolymerizing methacryloxypropyltrimethoxysilane with ethyl acrylate in the presence of a mercaptan. The liquid copolymers, mixed with titanate condensation catalysts, cross-linked at room temperature in the presence of atmospheric moisture to flexible cured elastomers.[45]

### 2.4.3. Amines

**2.4.3.1. Preparation.** Aminopropyl silanes may be prepared by adding trialkoxysilanes to allylamine,[46] but since allylamine is highly toxic, it is more convenient to prepare such compounds by hydrogenation of cyanoethylsilanes[47] or by reaction of chloropropyltrimethoxysilane with ammonia or amines[16]:

$$(EtO)_3SiCH_2CH_2CN + H_2 \rightarrow (EtO)_3SiCH_2CH_2CH_2NH_2$$

Aminophenyltrimethoxysilane was prepared from bromophenyl-trimethoxy silane and excess ammonia in presence of Cu in $Cu_2Cl_2$ catalysis[48]:

$$(MeO)_3SiC_6H_4Br + NH_3 \xrightarrow[110°]{Cu+CuCl_2} (MeO)_3SiC_6H_4NH_2 + NH_4Br$$

Aqueous solutions of nitroarylsilanolates were reduced by alkaline hydrazine in the presence of Raney nickel to obtain aminoarylsiliconates[49]:

$$NaO_{1.5}SiC_6H_4NO_2 + H_2NNH_2 \xrightarrow{NI} NaO_{1.5}SiC_6H_4NH_2 + N_2$$

Although aromatic silanes generally have better thermal stability than aliphatic silanes, the aminophenyl silanes showed little advantage over aminoalkyl silanes as coupling agents for high-temperature resins.

**2.4.3.2.   Reactions.** Chemical reaction is postulated between organo-functional groups of silylated surfaces and matrix resin in preparation of reinforced composites. The aminofunctional silanes, therefore, have shown promise as coupling agents with virtually all condensation-thermosetting polymers, e.g., epoxides, phenolics, melamines, furanes, isocyanates, etc., but not with unsaturated polyesters.

In addition, modified amino-organofunctional silanes are readily prepared from APS and AEAPS with a host of organic reagents.

Primary aminofunctions of silanes add exothermally at room temperature to activated double bonds of acrylic monomers to form functional-alkylated amines:

$$RNH_2 + CH_2{=}CHCOOR'X \;\rightarrow\; RNHCH_2CH_2COOR'X$$

Addition of amines to the double bonds of methacrylates, maleates, and fumarates is much slower and may require several hours at 100°C for complete reaction.[39] Adducts of APS and AEAPS to unsaturated polyesters were obtained by heating for about 1 hr at 100°C. The oligomeric adducts were proposed as sizing on glass fibers for reinforcement of polyester resins.[50]

Aminofunctional silanes react normally with cyclic carboxylic anhydrides to form the corresponding amic acids. Conversion to an anionic derivative generally improves the compatibility of aminofunctional silanes with anionic polymer emulsions[51]:

$$(MeO)_3SiCH_2CH_2CH_2NH_2 + \text{maleic anhydride} \;\rightarrow$$

$$(MeO)_3SiCH_2CH_2CH_2NHC(O)CH{=}CHCOOH$$

Reactions of amino-organofunctional silanes with reactive monomers are monitored readily by thin-layer chromatography on silica. A spot of unmodified amino-organofunctional silane deposited on silica and eluted with methanol shows very little movement, as indicated by dilute permagnagate or pH indicator. A mixture of unreacted monomer with amino silane show two spots: an immobile silane and the eluted monomer. After reaction is complete, there is only an immobile spot that may be modified in reactivity, e.g., an acrylic acid adduct may show an immobile spot at an acid pH in place of the original alkaline pH of the unmodified silane.

The effect of modifying the amine function of a silane is shown in primer data for PVC plastisols (Table 2.3). In this work, AEAPS was modified with equimolar amounts of readily available commercial monomers and applied from 25% solution in methanol to glass microscope

**Table 2.3.** Adhesion of PVC Plastisols to Primed Glass (Fused 2 min at 180°C)

| Silane (F) modifier in primer | Strength | | Silane (F) modifier in primer | Peel strength | |
|---|---|---|---|---|---|
| | N/cm | (lb./in.) | | N/cm | (lb./in.) |
| (Unprimed) | 0.2 | (0.1) | Diacetone acrylamide | 17.2 | (9.9) |
| Unmodified silane | 4.4 | (2.5) | Isobutoxymethyl acrylamide | 11.5 | (6.6) |
| Methylacrylate | 7.7 | (4.4) | Vinylbenzylchloride | 3.9 | (2.2) |
| Butylacrylate | 7.0 | (4.0) | Dodecylbenzylchloride | 0.4 | (0.2) |
| Ethoxyethylacrylate | 11.4 | (6.5) | Epichlorohydrin | 7.7 | (4.4) |
| 2-Ethylhexylacrylate | 7.3 | (4.2) | Phenylglycidyl ether | 5.4 | (3.1) |
| Hydroxyethylacrylate | 14.0 | (8.0) | Succinic anhydride | 18.4 | (10.5) |
| Hydroxypropylacrylate | 13.5 | (7.7) | Glutaric anhydride | 13.5 | (7.7) |
| Acrylonitrile | 11.9 | (6.8) | Phthalic anhydride | 15.4 | (8.8) |
| Cyanoethylacrylate | 19.2 | (11.0) | Decenylsuccinic anhydride | 0.7 | (0.4) |
| Acrylic acid | 18.4 | (10.5) | Nadic methyl anhydride | 9.6 | (5.5) |
| Methacrylic acid | 16.6 | (9.5) | Trimellitic anhydride | 0.9 | (0.5) |
| Partial hydrolyzate[73] | 25.0 | (14.0) | | | |

slides. A PVC plastisol was fused against the primed surfaces for 2 min at 180° and the film examined for peel strength on glass. Most of the modified silane primers (acrylates and anhydrides) were prepared by simple mixing in methanol. Others (methacrylic acid, halides, epoxides) required up to a few hours refluxing to complete the reaction. In general, polar functional modification of the silane improved its performance as a primer. With other polymer systems, such a series of primers might show a complete different order of effectiveness.

Amino-organofunctional silanes in nonpolar solvents show normal reactivity of aliphatic amines, but in polar solvents like alcohols or water, the NH groups on the third carbon atom from silicon are virtually inert. Advantage is made of this retarded reactivity to prepare one-component silane-modified epoxy primers. A solution of AEAPS in an alcohol with three equivalents of an epoxy resin reacts within a few hours at room temperature, as shown by a loss of two epoxy equivalents. After several weeks, one epoxy equivalent remains, but as soon as solvent is removed (or if the mixture were originally in toluene solvent) the third epoxy group reacts and the product cross-links. The low reactivity of —NH on the third carbon atom is believed to be due to hydrogen bonding with silyl oxygen in an internal cyclic structure.

Methacrylamides may be prepared by reacting methacryloyl chloride with an aminofunctional silane in the presence of an acid acceptor.[52] Amides are also formed by heating aminofunctional silanes with esters. Oligomeric amides derived from APS and dimethylcarbonate are recommended as primers for organic polycarbonate resins.[53]

Substituted ureas are obtained by warming APS with urethane in the presence of dibutyltin oxide catalyst[54]:

$$(EtO)_3SiCH_2CH_2CH_2NH_2 + H_2NCOOEt \rightarrow$$

$$(EtO)_3SiCH_2CH_2CH_2NHCONH_2 + EtOH$$

The amino groups of silanes APS and AEAPS are alkylated by organic halides. Cationic silanes derived from reaction of vinylbenzyl chloride with AEAPS are effective coupling agents for virtually all thermosetting resins.[55] In addition, they are among the most effective coupling agents for thermoplastic resins.[37] Bonding with nonreactive thermoplastics like the polyolefins is believed to be through an interpenetrating polymer network at the interface.[40]

Poly(carboxymethyl)amino compositions with strong chelating capacities for polyvalent cations are formed by the reaction of AEAPS with 2–3 mol haloacetic acid in the presence of a base[56]:

$$-NHCH_2CH_2NH_2 + 3ClCH_2COONa \xrightarrow{NaOH}$$

$$-N(CH_2COONa)CH_2CH_2N(CH_2COONa)_2 + 3NaCl$$

When aminofunctional silanes are neutralized with carboxylic acids in aqueous solution and applied to mineral surfaces, a portion of the acid remains after drying as the acid salt or as the amide (if dried at elevated temperature).[57] When a higher perfluoroalkanoic acid is used for neutralization, the treated surface is both hydrophobic and oleophobic.[58]

### 2.4.4.  Epoxy

**2.4.4.1.  Preparation.** Epoxy-organofunctional silicon compounds are prepared by addition of silanes to unsaturated epoxides or by epoxidation of unsaturated silanes having internal double bonds.[59] Epoxidation of vinylsilanes with peracetic acid gives poor yields of the vinyloxide in a sluggish reaction:

$$(MeO)_3SiH + CH_2=CHCH_2OCH_2\overline{CH-CH_2O} \xrightarrow{Pt}$$

$$(MeO)_3Si(CH_2)_3OCH_2\overline{CHCH_2O}$$

Addition of silanes to monoepoxides of 1,3-dienes gave predominately the 1,4-adducts[60]:

$$R_3SiH + CH_2=CH\overline{CHCH_2O} \xrightarrow{Pt} R_3SiOCH_2CH=CH-CH_2$$

Oxetane-functional silanes are obtained by addition of silanes to unsaturated oxetanes[61]:

$$(MeO)_3SiH + CH_2=CHCH_2OCH_2\overline{C(C_2H_5)-CHOCH_2} \rightarrow$$

$$(MeO)_3Si(CH_2)_3OCH_2\overline{C(C_2H_5)-CHOCH_2}$$

**2.4.4.2.  Reactions.** Active hydrogen compounds add to the epoxy group of epoxyfunctional silanes to form the expected hydroxy substituted ethers. Acid or basic catalysts may be used in addition of carboxylic acids, phenols, alcohols, or water. No catalyst is needed for addition of aliphatic amines[59]:

$$\equiv Si(CH_2)_3OCH_2\overline{CHCH_2O} + C_6H_5OH \xrightarrow{R_4NOH}$$

$$\equiv Si(CH_2)_3OCH_2CH(OH)CH_2OC_6H_5$$

Reaction products of epoxyfunctional siloxanes and polyethylene glycols are surface-active compounds useful as foam control agents in polyurethanes.[62,63]

Sodium sulfite adds readily to epoxy compounds to produce hydroxyalkyl sufonates.[63] The sulfonate derived from an epoxy-functional silane (GPMS) is a very good stabilizer of sodium silicates solution[64]:

$$\equiv SiCH_2CH_2CH_2OCH_2CHCH_2O + Na_2SO_3 \longrightarrow$$

$$= SiCH_2CH_2CH_2OCH(OH)CH_2SO_3Na + NaOH$$

The major applications of epoxy-organofunctional silanes are as coupling agents in reinforced condensation-thermosetting polymers, like epoxies, phenolics, melamines, urethanes, etc. In applying epoxy-functional silanes from aqueous solution, the pH of the solution must be maintained above 4 to prevent hydrolysis of the epoxide to a glyceryl ether.

### 2.4.5. Mercaptans

Preparations of mercaptofunctional silanes from unsaturated silanes or chloroalkylsilanes were described in earlier sections. The mercaptofunctional silanes are convenient chain-growth regulators in vinyl polymerization and introduce a trimethoxysilane functionality into each polymer molecule by chain transfer. Additional silane functionality may be introduced by copolymerizing unsaturated silanes with the other monomers. About ten trimethoxysilane functions per polymer molecule are required for adhesion of thermoplastics to glass.[65]

Mercaptofunctional silanes add readily to acrylates, methacrylates, itaconates, fumarates, etc., in the presence of basic catalysts. This is one of the few reactions where methacrylates react as easily as acrylates[66]:

$$RSH + CH_2 = CR'COOH_3 \xrightarrow{\text{NaOMe}}$$

$$R'SCH_2 - CHRCOOCH_3 \qquad R' = Me \text{ or } H$$

Free-radical initiators are used to obtain addition of mercaptofunctional silanes to nonactivated double bonds as in SBR copolymers and other vulcanizable organic elastomers.[67]

Mercaptofunctional silanes are recommended as coupling agents on particulate mineral to upgrade them to reinforcing fillers in vulcanized rubber.[68]

### 2.4.6.  Carboxylic acids

Carboxyl-functional siliconates are most readily prepared by saponification of nitrile- or ester-functional silanes. Cyanoalkylsilanes are prepared by base-catalyzed addition of $HSiCl_3$ to acrylonitrile[69]:

$$Cl_3SiH + CH_2{=}CHCN \xrightarrow{Me_3N} Cl_3SiCH_2CH_2CN$$

$$Cl_3SiCH_2CH_2CN + 4Na_2OH \xrightarrow{H_2O}$$

$$O_{1.5}SiCH_2CH_2COONa + 3NaCl + NH_2{\uparrow}$$

A tolyl silicone derived by hydrolysis of $CH_3C_6H_4{-}SiCl_3$ was oxidized by alkaline permanganate to recover a soluble carboxyphenyl siliconate that could be separated from $MnO_2$ by filtration. Acidification of the filtrate gave a solid carboxyphenyl siloxanol oligomer that gave stable solutions in acetone/water[70]:

$$(NaO)OSiC_6H_4COONa \xrightarrow{acid} (HO)OSiC_6H_4COOH$$

Carboxy-modified organofunctional silanes were also prepared from mercapto- or aminofunctional silanes and unsaturated acids, or from aminofunctional silanes and cyclic anhydrides or chloroacetic acid, etc.

Since the carboxyfunctional silicones are generally applied from aqueous solutions, it is not necessary to isolate the free acid. Carboxyfunctional siliconates are very good coupling agents for epoxy resins, but have not been used commerically. Carboxyphenyl siliconate has extremely good heat stability and is recommended in polybenzimidazole composities, but not with polyimides.

Carboxyalkylsiliconates have high water solubility and act as solution stabilizers for silicate corrosion inhibitors (see Chapter 3).

Acid-functional silanes have been applied as coupling agents on glass or fillers with acid-functional polymeric film formers in the presence of metal cations to obtain ionomer structures at the interface.[71] By matching compatibilities of the film former with neutral matrix polymers (e.g., polyolefins, nylon, etc.) a remarkably strong, water-resistant bond was formed with the filler. Optimum water resistance was observed when the acid-functional materials were modified with sodium or zinc ions sufficient to neutralize 30–40% of the acid groups. The ionomer structure at the interface is mobile at high temperature and high shear rates (extrusion or injection molding), but acts like a crosslinked structure when the composite cools.

### 2.4.7. Alcohols

Direct addition of silanes to unsaturated alcohols in the presence of platinum catalyst is complicated by competing solvolysis of the silane hydrogen[72]:

$$(MeO)_3SiH + CH_2{=}CHCH_2OH \xrightarrow{Pt} (MeO)_3SiOCH_2CH{=}CH_2 + H_2$$

Addition to unsaturated secondary alcohols or phenols proceeds readily to form the double-bond adduct.[73]

$$(EtO)_3SiH + o\text{-}CH_2{=}CHCH_2C_6H_4OH \xrightarrow{Pt}$$

$$o\text{-}[(EtO)_3SiCH_2CH_2CH_2]C_6H_4OH$$

Adducts of silanes to allyl esters may be alcoholized to recover hydroxypropylsilanes.

$$(MeO)_3SiH + CH_2{=}CHCH_2OCOCH_3 \xrightarrow{Pt}$$

$$(MeO)_3SiCH_2CH_2CH_2OCOCH_3 \quad (I)$$

$$(I) + CH_3OH \xrightarrow{H+} (MeO)_3SiCH_2CH_2CH_2OH + MeOAc$$

Hydroxyethyl groups are cleaved readily from silicon; hydroxymethyl silanes cleave in strong alkali; but hydroxypropyl groups on silicon have the normal stability and reactivity of aliphatic alcohols.

Hydroxylfunctional silanes are also prepared from mercaptofunctional silanes and unsaturated alcohols, from aminofunctional silanes with epoxides or with hydroxyfunctional acrylates, or from epoxyfunctional silanes and glycols or water. One of the simplest methods of obtaining hydroxyfunctional silanes from commercial materials is to dissolve an epoxyfunctional silane in water at a pH of 3 or less:

$$(HO)OSiCH_2CH_2CH_2OCH_2\overline{CHCH_2}O + H_2O \xrightarrow{H+}$$

$$(HO)OSiCH_2CH_2CH_2OCH_2CH(OH)CH_2OH$$

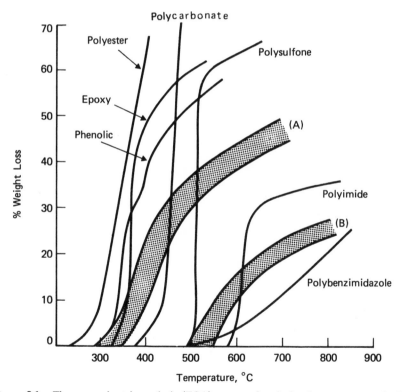

**Figure 2.1.** Thermogravimetric analysis (TGA) curves of typical polymers compared with silane coupling agents: (A) Aliphatic organofunctional silanes; (B) Aromatic organofunctional silanes.

### 2.4.8. Aromatic Silanes

Aromatic silanes, in general, have much better heat stability and oxidation resistance than aliphatic silanes, and have been proposed as a general class of coupling agents for high-temperature resins (see Figure 2.1).[74] To function as coupling agents with thermosetting resins, however, silanes must have a high degree of reactivity with the resin during early stages of cure, a quality simple aromatic silanes lack.

Carboxyphenylsilanes have been recommended with polybenzimidazole resins, while chloromethylphenyl silanes showed most promise as coupling agents for polyimide resins.

The presence of an aromatic ring in an aralkyl silane does not always contribute significant improvement in oxidation resistance over comparable apliphatic silanes because the ultimate bond to silicon is an aliphatic carbon. From isothermal weight loss studies in air at 300°C it was discovered that a vinylbenzyl-substituted aminofunctional silane had much better stability than the parent amino-functional silane (Figure 2.2). Polymerization of the styrene function provides additional cross-linking that apparently helps to stabilize the entire structure. Mixtures of phenyltrimethoxysilane with aminofunctional silanes have been proposed as coupling agents with improved water resistance and thermal stability.[5] The cohydrolyzed silanes

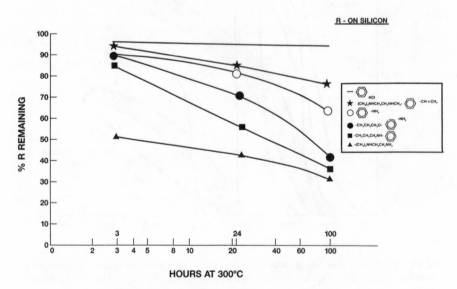

**Figure 2.2.** Stability of $RSiO_{1.5}$ in air at 300°C.

deposit a mixed silicone resin on the surface that has much of the thermal stability of phenylsilicones, while retaining adequate reactivity of the amino-organofunctional silane (Figure 2.2).

## References

1. H. A. Clark and E. P. Plueddemann. *Mod. Plast.* **40**(6), 133 (1963).
2. P. W. Erickson and I. Silver. U.S. Patent 2,776,910 (to U.S. Navy) (1955).
3. E. P. Plueddemann. *J. Paint Technol.* **40**(516), 1 (1968).
4. C. Eaborn. *Organosilicon Compounds,* Butterworth, London (1960).
5. W. Noll. *Chemistry and Technology of Silicones,* Academic, New York (1968).
6. J. L. Speier. U.S. Patent 2,723, 987 (to Dow Corning) (1955).
7. L. M. Shorr. *J. Am. Chem. Soc.* **76**, 1390 (154).
8. E. P. Plueddemann and G. Fanger. *J. Am. Chem. Soc.* **81**, 2635 (1959).
9. E. R. Pohl. SPI, 38th Ann. Tech. Conf. Reinf. Plast. 4-B (1983).
10. P. Dreyfuss, L. J. Fetters, and A. N. Gent. *Macromolecules* **11**(51), 1036 (1978).
11. R. L. Kaas and J. L. Kardos. SPE, 32nd ANTEC, Paper 22 (1976).
12. J. L. Speier. *J. A. Chem. Soc.* **73**, 824 (1951); U.S. Patent 2,510,148 (to Dow Corning) (1950).
13. L. H. Sommer and F. C. Whitmore. *J. Am. Chem. Soc.* **68**, 485 (1946).
14. B. Arkles and W. R. Peterson. SPI, 35th Ann. Tech. Conf. Reinf. Plast. 20-A (1980).
15. E. P. Plueddenmann. SPI, 21st Ann. Tech. Conf. Reinf. Plast. 3-D (1966).
16. J. L. Speier, C. A. Roth, and J. W. Ryan. *J. Org. Chem.* **36**(21), 3120 (1971).
17. A. J. Isquith, E. A. Abbott, and P. A. Walters. *Appl. Microbiol.* **24**(6), 859 (1972).
18. E. P. Plueddemann. U.S. Patent 3,427, 340 (to Dow Corning) (1969).
19. R. L. Merker. U.S. Patent 2,793,223 (to Dow Corning) (1957).
20. E. P. Plueddemann. U.S. Patent 4,093,641 (to Dow Corning) (1978).
21. E. P. Plueddemann. U.S. Patent applied for (to Dow Corning) (1989).
22. J. L. Speier. U.S. Patent 4,082,790 (to Dow Corning) (1978).
23. G. M. Omietanski and H. E. Petty. U.S. Patent 3,849,471 (to U.C.C.) (1974).
24. G. E. LeGrow. U.S. Patent 3,631,194 (to Dow Corning) (1971).
25. J. L. Speier. U.S. Patent 2,510,146 (to Dow Corning) (1950).
26. A. Berger. U.S. Patent 3,821,218 (to General Electric) (1975).
27. J. B. Thompson. U.S. Patent 3,697,551 (to Hercules, Inc.) (1972).
28. L. H. Sommer, R. E. Van Strein, and F. C. Whitmore. *J. Am. Chem. Soc.* **71**, 3056 (1949).
29. C. A. Burkhard and R. H. Krieble. *J. Am. Chem. Soc.* **69**, 2687 (1947).
30. D. L. Bailey and A. N. Pines. *Ind. Eng. Chem.* **46**, 2363 (1954).
31. A. A. Oswald and L. L. Murrell. U.S. Patent 4,081,803 (to Exxon) (1978).
32. E. P. Plueddemann. U.S. Patent 3,079, 361 (to Dow Corning) (1963).
33. E. P. Plueddemann, H. A. Clark, L. E. Nelson, and K. R. Hoffman. *Mod. Plast.* **39**, 135 (1962).
34. E. P. Plueddemann and H. A. Clark. U.S. Patent 3,258,477 (to Dow Corning) (1966).
35. R. H. Krahnke, K. W. Michael, and E. P. Plueddemann. U.S. Patent 3,631,085 (to Dow Corning) (1971).
36. E. P. Plueddemann. U.S. Patent 3,734,763 (to Dow Corning) (1973).
37. E. P. Plueddemann. U.S. Patent 3,819,675 (to Dow Corning) (1974).
38. Freeport Kaolin Co. *Translink 37, Tech. Bull.* (1969).
39. E. P. Plueddemann. SPI, 27th Ann. Tech. Conf. Reinf. Plast. 11-B (1972).
40. E. P. Plueddemann and G. L. Stark. SPI, 35th Ann. Tech. Conf. Reinf. Plast. 20-C (1980).

41. H. G. Scott. U.S. Patent 3,646,155 (to Midland Silicones) (1972).
42. R. P. Louthan. U.S. Patent 3,890,213 (to Phillips Petroleum Co.) (1975).
43. E. P. Plueddemann. U.S. Patent 3,317,461 (to Dow Corning) (1967).
44. R. Bruening, H. Hanisch, and H. J. Hass. Ger. Patent 2,730,008 (to Dynamit Nobel) (1979).
45. E. P. Plueddemann. U.S. Patent 3,453,230 (to Dow Corning) (1969).
46. S. Reichel. U.S. Patent 3,665,027 (to Inst. Silikon-u Fluokarbon Chem.) (1972).
47. L.H. Sommer and J. Rockett. *J. Am. Chem. Soc.* **73**, 5130 (1951).
48. E. W. Bennett and P. Orenski. U.S. Patent 3,646,087 (to U.C.C. (1972).
49. E. P. Plueddemann and C. A. Roth. U.S. Patent 3,509,196 (to Dow Corning) (1970).
50. E. J. Peppe and J. G. Marsden. U.S. Patent 4,122,074 (to U.C.C. (1978).
51. E. P. Plueddemann. U.S. Patent 3,956,353 (to Dow Corning) (1976).
52. T. A. TeGrotenhuis. U.S. Patent 3,249,461 (1966).
53. D. G. LeGrand and G. G. Vitale. U.S. Patent 4,040,882 (to General Electric), (1977).
54. E. J. Peppe and J. G. Marsden. U.S. Patent 3,671,562 (to U.C.C.) (1972).
55. E. P. Plueddemann. SPI, 27th Ann. Tech. Conf. Reinf. Plast. 21-B (1972).
56. E. P. Plueddemann. U.S. Patent 4,071,546 (to Dow Corning) (1978).
57. E. P. Plueddemann and G. L. Stark, *Mod. Plast.* **54**(8), 76 (1977).
58. E. P. Plueddemann. U.S. Patent 4,687,057 (to Dow Corning) (1986).
59. E. P. Plueddemann and G. Fanger. *J. Am. Chem. Soc.* **81**, 2632 (1959).
60. V. M. Al'bitskaya, I. E. Sharikova, and A. A. Petrov. *Zh. Obschch. Khim.* **33**(11), 3770 (1963).
61. E. P. Plueddemann. U.S. Patent 3,338,867 (to Dow Corning) (1967).
62. E. P. Plueddemann. U.S. Patent 3,057,901 (to Dow Corning) (1962).
63. C. M. Suter and H. B. Milne. *J. Am. Chem. Soc.* **65**, 582 (1943).
64. E. P. Plueddemann. U. S. Patent 4,503,242 (to Dow Corning) (1985).
65. E. P. Plueddemann. SPI, 20th Ann. Tech. Conf. Reinf. Plast. 19-A (1965).
66. C. D. Hurd and L. L. Gershbein. *J. Am. Chem. Soc.* **69**, 2328 (1947).
67. J. L. Speier and J. W. Keil. U.S. Patent 3,440,302 (to Dow Corning) (1969).
68. S. Nozakura and S. Konotsune. *Bull. Chem. Soc. Jpn.* **29**, 322 (1956).
69. A. N. Pines and E. A. Zientek. U.S. Patent 3,198,820 (to U.C.C.) (1965).
70. R. Wong and J. C. Hood, AFML-TR-65-316 Summary Tech. Report (July 1965).
71. E. P. Plueddemann, *J. Adhesion Sci. Tech.* **3**(2), 131 (1988).
72. J. L. Speier, J. A. Wolster, and G. H. Barnes. *J. Am. Chem. Soc.* **79**, 974 (1957).
73. E. P. Plueddemann. U.S. Patent 3,328,450 (to Dow Corning) (1967).
74. E. P. Plueddemann. SPI, 22nd Ann. Tech. Conf. Reinf. Plast. 9A (1967).
75. E. P. Plueddemann and P. G. Pape. SPI, 40th Ann. Tech. Conf. Reinf. Plast. 17-F (1985).

# 3 | Aqueous Solutions of Silane Coupling Agents

## 3.1. Introduction

Because organofunctional alkoxysilanes are often hydrolyzed before being applied to surfaces to function as coupling agents, it is important to understand their reactions both with and in water. Commercial practice is to apply silane coupling agents to glass from aqueous solutions of the alkoxysilanes. Organofunctional trialkoxysilanes hydrolyze in water and then condense to oligomeric siloxanols as described in Section 2.3.3. The stability and reactivity of aqueous solutions of silanes depend on many factors, including the nature of the organofunctional group on silicon. Because compounds with neutral organofunctional groups behave differently from those with cationic or anionic functions, they will be discussed separately.

## 3.2. Alkoxysilanes with Neutral Organofunctional Groups

### 3.2.1. Solubility in Water

A few organotrialkoxysilanes are immediately miscible with water, but the majority do not become soluble until the alkoxy groups hydrolyze. Hydrolysis of the alkoxy groups, however, requires molecular contact of water with the alkoxysilane. For this reason, it is very difficult to hydrolyze some alkoxysilanes directly, and it is indeed a very slow process to hydrolyze alkoxysilanes dissolved in hydrocarbon solvents. Silane coupling agents may be dissolved in water-insoluble plasticizers and stirred into aqueous emulsions of polymers to get stable compositions. The silane remains as an unhydrolyzed monomer during storage so it has maximum effectiveness as an adhesion promoter when the latex is applied to a surface and dried.

Organotrialkoxysilanes are commonly dissolved in water by shaking or stirring vigorously with acidified water until a clear solution results. Finer dispersion of the alkoxysilane with the aid of an emulsifying agent gives more rapid hydrolysis to clear solutions. At a pH of 3–5, dilute solutions are stable for more than a day and consist predominantly of the monomeric silane triol.

The kinetics and mechanisms of acid- and base-catalyzed hydrolysis of alkyltrialkoxysilanes in aqueous solution were reported by Pohl.[1] Acid-catalyzed hydrolysis involved attack on an alkoxy oxygen by a hydronium ion followed by a bimolecular $S_N2$-type displacement of the leaving group by water. Alkali-catalyzed hydrolysis involved attack on silicon by a hydroxyl ion to form a pentacoordinate intermediate followed by a bimolecular displacement of alkoxyl by hydroxyl. Rates of hydrolysis by both mechanisms were influenced by the nature of the alkyl group on silicon as well as the leaving alkoxyl group. Half-lives of various alkyltrialkoxy silanes in acid and alkaline aqueous solutions are shown in Table 3.1.

The pH profile for the first hydrolysis step of γ-glycidoxypropyltria-methoxysilane (GPS) is shown in Figure 3.1.[2] NMR spectra of dilute aqueous solutions of GPS show no significant concentration of intermediate methoxysilanols, but only the silanetriol and its condensed siloxane oligomer. Apparently the second and third alkoxy groups hydrolyze faster

**Table 3.1.** Hydrolysis of Alkyltrialkoxysilanes[a]

| R-Group of $R\text{-Si}(OCH_2CH_2OCH_3)_3$ | pH 9 (min.) | pH 5 (min.) |
|---|---|---|
| $ClCH_2-$ | 0.04 | 15 |
| $CH_2{=}CH-$ | 1.8 | 13 |
| $CH_3-$ | 6.8 | 21 |
| $C_2H_5-$ | 25 | 31 |
| $C_6H_5-$ | 14 | 33 |
| $CH_3CH_2CH_2-$ | 58 | 55 |
| cyclo-$C_6H-$ | 340 | 165 |

| R'-Group of $CH_3O(CH_2CH_2O)_3CH_2CH_2Si(OR')_3$ | | |
|---|---|---|
| $CH_3-$ | | 4.3 |
| $CH_3CH_2-$ | | 23 |
| $CH_3OCH_2CH_2-$ | | 41 |

[a] Half-lives in acid and alkaline solutions.

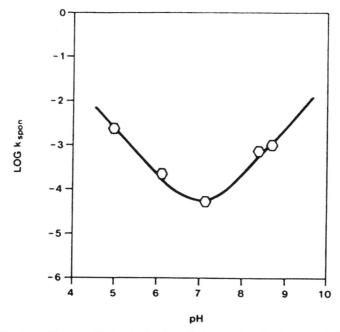

**Figure 3.1.** The pH-rate profile for the first hydrolysis step of $\gamma$-glycidoxypropyltrimethoxysilane.

than the first. The silane triol content can be monitored uniquely by laser Raman spectroscopy due to its strong characteristic absorption mode.[3] This work indicates that very dilute solutions remain as silanetriols, but there is a particular concentration at which the triols form aggregated monomers through hydrogen bonding. This transition for vinylsilanetriol is about 1% concentration in water and for $\gamma$-aminopropylsilane triol at about 0.15%.

### 3.2.2. Solution Stability

Solubility of hydrolyzates containing silane triols, $RSi(OH)_3$, decreases as silanols condense to oligomeric siloxanols. Clear aqueous solutions of hydrolyzed silanes may become hazy upon standing and eventually separate as oily droplets. In practical applications, the hydrolyzate has lost its usefulness at the point of haze formation. The stability of aqueous silanol solutions was studied by comparing 100 ml of 3% aqueous silane solution at pH 2 in clear glass 4-oz. square bottles.[4]

Solutions were observed periodically for clarity. Stability limit was taken at the time when a printed page could not be read through the bottle because of the haze. This endpoint was quite sharp; as solutions approached the stability limit, the haze generally developed in a short time.

Stabilities of dilute aqueous solutions of a few organotrimethoxysilanes in water at pH 2 are shown in Table 3.2. Differences in time to haze are not a good indication of different rates of condensation of silanols, but rather a measure of relative solubility in water of intermediate siloxanols.

Pohl[5] and Osterholz studied the rate of condensation of silanols in deuterium oxide and found that two silane triols had similar pH profiles, but that they differed from a diorganosilane diol or a triorganosilanol (Figure 3.2). These data confirm many practical observations that silane coupling agents (triols) have a maximum solution stability in water at a pH of about 4.

The compositions of vinylsiloxanols or vinylsiliconates in water were determined by the Lentz technique[6] as summarized in Table 3.3. In this technique, the aqueous solutions are treated under mild conditions with acidic trimethylsilylating reagents that convert silanols and silanolate ions to stable trimethylsilyl endblocked siloxanes. Lower members of this series are readily separated and estimated by gas–liquid phase chromatography.

Fresh solutions were observed to be predominantly monomer triol. After several hours at pH 2, as the solution became hazy there still remained about 34% monomer, but dimer, trimer, and higher oligomers were identified.

### 3.2.3. Siloxane Formation

Molecular weights of precipitated or extracted siloxanes may be estimated by gel permeation chromatography (GPC). The precipitated siloxane derived from $\gamma$-methacryloxylpropyl-trimethoxysilane ($\gamma$-MPS) after drying at room temperature showed a broad molecular weight range with a peak at about 8,000. The same siloxane extracted from treated and dried clay filler had a much narrower molecular weight range with a peak at about

**Table 3.2.** Stability of 3% Aqueous Solutions of $RSi(OMe)_3$ at pH 2

| R-Group of $RSi(OMe)_3$ | Time to haze, hr |
|---|---|
| Me | 10.0 |
| Vi | 5.5 |
| Et | 7.0 |
| Pr | 5.25 |
| $CF_3CH_2CH_2-$ | 14.0 |
| $ClCH_2CH_2CH_2-$ | 5.2 |
| $n$-Bu | 0.75 |
| Mixed amyl | 5.75 |
| Ph | 7.0 |

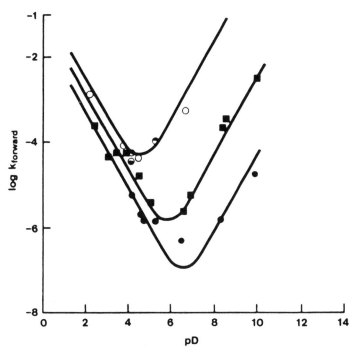

**Figure 3.2.** The pH-rate profile for the condensation of (○) γ-glycidoxypropylsilanetriol, (◓) γ-methacryloxypropylsilanetriol, (◐) γ-aminopropylsilanetriol, (■) γ-glycidoxypropyl-methylsilanediol, and (●) γ-aminopropyldimethylsilanol to their respective disiloxanes.

**Table 3.3.** Silanol(ate) Composition (Lentz Technique) of 10% ViSi(OMe)$_3$ Solutions in Water

| Solution | Composition, % | | | |
|---|---|---|---|---|
| | Monomer | Dimer | Trimer | Tetramer |
| Fresh, pH 2.0 | 82 | 15 | 3 | — |
| Hazy, pH 2.0 | 34 | 23 | 30 | 13 |
| 1 : 1 NaOH | 45 | 31 | 20 | 4 |
| 2 : 1 NaOH | 78 | 17 | 5 | — |
| 3 : 1 NaOH | 93 | 7 | — | — |

900, equivalent to a tetramer.[7] Siloxane formation on a filler surface apparently was modified by reactions of silanols with the filler surface.

Condensation of silanols in solution may be monitored by the $^{13}C$ chemical shifts of a methyl-on-silicon by nuclear magnetic resonance (NMR) spectroscopy.[5] NMR spectroscopy of $\gamma$-glycidoxypropylmethyl-silane diol in deuterium oxide showed that the condensation and rehydrolysis were in true equilibrium with an equilibrium constant of condensation $K = 210$ (Figure 3.3). There was little evidence of any condensation of the second silanol during 90 min of this test.

Many studies have been reported on the composition of silicone resin intermediates derived from hydrolysis of reactive silanes, $RSiX_3$. Although these resin intermediates are not water soluble, they indicate a possible direction of condensation of hydrolyzed silane coupling agents in organic solvents. Coupling agents with strongly hydrophilic organofunctional groups could retain water solubility even after condensation of silanols to siloxanes is complete.

Andrianov and Izmaylov[8] observed that products from hydrolytic poly-condensation of higher alkyltrichlorosilanes are soluble in organic solvents

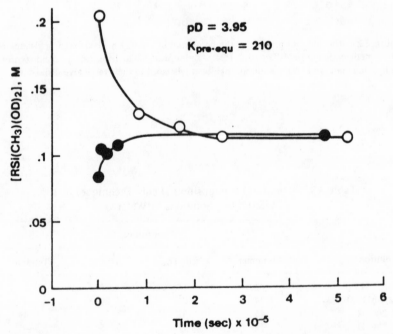

**Figure 3.3.** Concentration of $\gamma$-glycidoxypropylmethylsilanediol vs. time. (O) silanediol, (●) disiloxane.

and may be distilled under vacuum, while hydrolyzates of lower alkyltrichlorosilanes condense rapidly to infusible cross-linked resins. With increasing size of alkyl group on silicon, there is a greater stability of intermediate silanols and a greater tendency for cyclic siloxane formation upon condensation. The ultimate distilled products were cubical octamers, or silsesquioxanes.

Sprung and Gunther[9] studied the hydrolysis of *n*-amyltriethoxysilane and phenyltriethoxysilane in homogeneous solution with both acid and base catalysts. Acid hydrolysis gave soluble reactive oligomers with residual silanol and ethoxysilane end groups. Alkaline hydrolysis gave soluble oligomers with much lower silanol and ethoxysilane content. Both types of oligomers were postulated to consist of fused cyclic tetrasiloxanes ranging from "ladderlike" structures to cubical octamers.

Cyclohexylsilanetriol and phenylsilanetriol are readily isolated as pure monomers. Brown and Vogt[10] studied the polycondensation of these triols. The condensation products obtained from phenylsilanetriol may be taken as typical of products obtained by hydrolysis and condensation of organofunctional silane coupling agents. Condensation of silanols in toluene solution leads to predominantly cyclic tetrasiloxanes. Very few cyclic trisiloxanes are obtained except by high-temperature alkali-catalyzed rearrangement. Using $T(OH)_3$ or $\curlywedge$ as symbols for silane triols, the predominant course of condensation from solution is outlined in Scheme I.

At first glance, it might appear surprising that a polycondensation follows such a simple pathway, but functional groups variously attached to branchable chains and cyclized structures (unlike those attached to the ends of straight chains) are grossly nonequivalent structurally, and hence can have very different reactivity. Moreover, much of the condensation must proceed via intramolecular cyclization rather than intermolecular reactions, and hence will be guided by the specific cyclization tendencies of the ring systems, rather than by the statistics of acyclic combination.

Two sources of reactivity differences may be identified. One is steric hindrance, which favors straight (rather than branched) chains in early stages of reaction and cyclization rather than further chain growth whenever possible.

A second source of selectivity is the strong multiple hydrogen bonding which can occur when two cyclic polysilanols are brought together. Phenylsilanetriol and its intermediate condensation products have surfactantlike structures with silanol groups at one end and phenyl groups at the other. Solutions of such species in nonpolar solvents should form micelles having the silanol groups on the inside united by hydrogen bonding. If siloxane formation proceeded within the micelle, a compact polycyclic would result with as much of the micelle form as could be retained strainlessly.

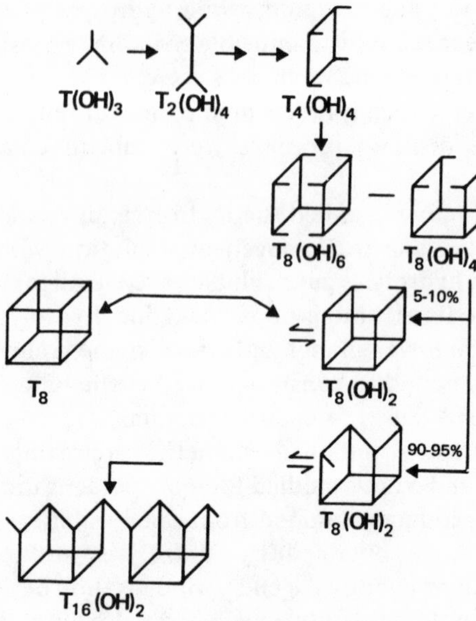

**Scheme 1**

Silane triols may follow a somewhat different condensation path in aqueous solution. Silanol groups will be hydrogen-bonded to water and to organofunctional groups as well as to one another. Any micelle structure will be inverted with hydrophilic silanol groups at the water interface. Silanol groups will be relatively isolated from one another and stabilized by hydrogen bonding to water. It is expected, therefore, that dilute aqueous solutions of hydrophilic organofunctional silanes will retain much silanetriol, or low-molecular oligomeric siloxanol structure, especially polycyclic structures resembling the silicone resin intermediates described by Brown.[11]

The amount of silanes deposited on mineral surfaces from water increases rapidly as concentations increase to certain transition concentrations; above this concentration silane uptake increases more slowly. Transitions of this kind may be attributed to the onset of micelle formation of aqueous silanetriols. Transition concentrations reported by Ishida et al.[12] for vinyltrimethoxysilane were 1.0% by weight, γ-MPS at 0.4%, and γ-APS at 0.15% by weight in water.

### 3.2.4. Solution in Aqueous Alcohols

Partial hydrolysis of tetraethoxysilane in alcohol solvents was studied by GLC determination of alcohols, and NMR examination of the devolatilized residue was used to determine silanol and alkoxy groups remaining on silicon.[13] Alkoxy exchange accompanied the hydrolysis, but was observed only after hydrolysis was far advanced, suggesting that silanol groups were reesterified by the alcohols. Hydrolysis in $\beta$-deuterated ethanol resulted in equal distribution of ethanols between polymer and solvent, indicating true equilibrium conditions of hydrolysis and reesterification.

Primers are commonly formulated from silane coupling agents by partially hydrolyzing the alkoxy silanes in alcohols (see Section 6.6). An apparent equilibrium is reached of alkoxysilanes and siloxane oligomers along with alkoxy exchange products. A convenient way to prepare a primer is to add about 10% water to a silane dissolved in a suitable alcohol.[14] Equilibrium conditions are reached in a few hours, after which the primer is essentially unchanged for over a year.

## 3.3. Cationic-Organofunctional Silanes

### 3.3.1. Amino Organofunctional Trialkoxysilanes

Neutral organofunctional silanes are hydrolyzed in aqueous solutions at pH 4 to obtain silane triols with moderate stability in water. Except for very dilute solutions, these silane triols condense to siloxane that gel or precipitate oily oligomers within a few hours. When amino-organofunctional silanes are hydrolyzed in excess acetic acid at pH 4, silane triols again are formed and solutions up to 50% in water remain stable indefinitely due to the strong hydrophilic nature of the organofunctional group.

When aqueous amino-organofunctional silane solutions are prepared at their natural pH, the products immediately form oligomers that retain solubility in water to moderate concentrations. Since solutions of neutral organic siloxanols are rapidly precipitated as insoluble gels by addition of amines, it is rather surprising to observe that amino-organofunctional trialkoxysilanes are readily soluble in water to give solutions of unlimited stability at a pH that causes rapid condensation of silanols.

The chemical structure of the silane in solution is now fairly well understood. It appears from $^1$HNMR examination of aqueous solutions of $(MeO)_3SiCH_2CH_2CH_2NHCH_2CH_2NH$ that the alkoxy groups are hydrolyzed almost immediately in water at alkaline pH. Attempts to extract volatile siloxanes from concentrated aqueous solutions of aminoalkylsiloxanes by the Lentz technique were not successful, suggesting that no monomeric silanol exited in solution. Either the products are too highly condensed to give a volatile product, or silanols are in a form that do not give trimethylsiloxanes with trimethylsilylating agents.

Commercial amino-organofunctional silanes have an amine function on the third carbon relative to silicon. It was at one time proposed that the unique solubility of these materials might be due to internal cyclization between nitrogen and silicon or silanolate ion.[4] Some reaction between silicon and nitrogen on the third carbon atom would be expected from analogy with pentacoordinate bridgehead siloxazoladines described by Frye et al.[15]

Alcoholysis of triethanolamine led exclusively to crystalline monomeric structures instead of polymeric structures that might reasonably have been expected from condensation of such highly functional reactants:

$$ZSi(OR)_3 + (HOCH_2CH_2)_3N \rightarrow ZSi(OCH_2CH_2)_3N + 3ROH$$

These silanes were believed to have transannular dative bonding between the nitrogen and silicon atoms, as deduced from reduced reactivity observed for both nitrogen and silicon:

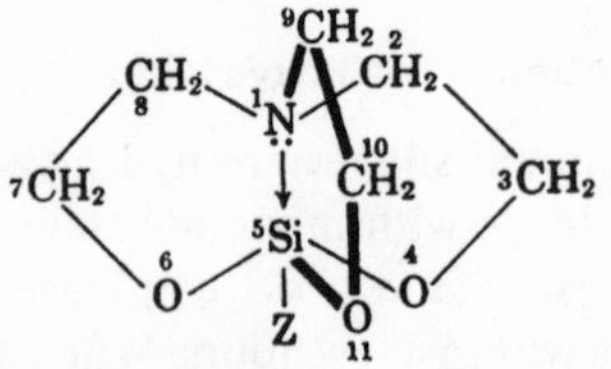

Triptych-siloxazolidines are neutralized only slowly by perchloric acid in glacial acetic acid. The stretching frequency for Si—H (when Z=H) showed clearly as increased electron supply to silicon, displacing the infrared absorption to lower-energy regions.

No evidence for such structures could be observed in aminopropylsilane solutions by $^1$H NMR or Fourier transform IR techniques. At the normal alkaline pH of these solutions there is no evidence of amine salt formation or pentacoordinate silicon, but a most likely structure is an internal hydrogen-bonded seven-membered ring.

Of three possible conformations proposed by Chiang et al.,[16] the structure shown in Figure 3.4 is consistent with the chemistry of aminopropylsilanes. Bonding of a silanol hydrogen explains the stability against cross-linking of trifunctional aminopropylsilanes in relatively concentrated aqueous solutions. Bonding of an amine hydrogen explains the lack of activity of aminopropyl silanes (e.g., with epoxies) in alcoholic solvents.

In acid solutions, silanol groups are relatively stable, while amines are protonated to soluble salt structures. Since silanol groups have an isoelectric point at about 2–3, these silanes in water in the pH range of 3–6 must exist

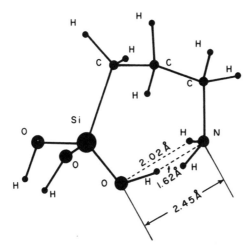

**Figure 3.4.** Possible conformation of aminopropylilane triol with total conformational energy of −26.3 kcal/mol.

as zwitterions of structures like

$$^-O(HO)_2Si-CH_2CH_2CH_2NH_3^{\oplus}$$

In mildly alkaline solution, silanol groups are much less stable and condense to polysiloxanes. Bulky substituted alkyl groups promote cyclization during condensation, so that low-molecular oligomeric silsesquioxanes (cage structures) result. If the aminoalkyl group on silicon is sufficiently hydrophilic, the polysiloxanes retain water solubility. It is probable that structures of the alkylsilane oligomers in water are in equilibrium. An oligomeric partial hydrolyzate, when diluted to 0.5% in water at pH 4 was not nearly as effective a coupling agent as a 0.5% solution of silane triol applied to glass. After aging for 8 days or warming for 1 hr to 90°C, the 0.5% oligomer solution approached the performance of the silane triol.[14] In dilute solutions, more open siloxanol structures predominate, while in concentrated solutions low-molecular-weight cage structures are favored.

Laser Raman spectroscopy was used by Ishida et al.[12] to determine silanol concentration of dilute aqueous solutions of γ-aminopropyltriethoxysilane (APS). Silanetriol was observed as a line at $712 \, \text{cm}^{-1}$ and compared with the $881 \, \text{cm}^{-1}$ line of ethanol, which acted as an internal standard. The relative intensity $I_{712}/I_{881}$ was plotted against concentration of APS in water (Figure 3.5). The proportion of monomeric silanetriol increased gradually with decreasing solution concentration to about 1% by weight. Further decrease in concentration of APS gave a rapid increase in silanetriol content.

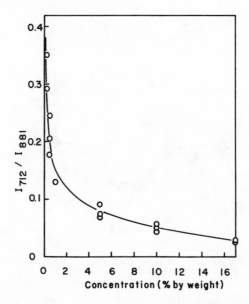

**Figure 3.5.** Relative Raman intensity $I_{712}/I_{881}$ vs. the concentration of silane solutions obtained for γ-APS in water.

This curve is compared with data of Johannson et al.[17] on pickup of silane by glass from dilute aqueous solutions of [14]C-labeled APS (Figure 3.6). The two curves are almost mirror images of each other, reflecting rapid increase in absorption of monomeric silanetriol with increasing concentration of APS to about 0.15%, which must correspond to the breakpoint for silanetriol in solution.

When APS was neutralized with excess acetic acid before dissolving in water, fresh solutions at 1–3% concentration contained high proportions of silanetriol. Exactly neutralized APS hydrolyzed very slowly in water at pH 7. Under these conditions, its solutions resembled those of VTS or MPS.

Under strongly alkaline conditions, aqueous solutions of aminoalkyl-silanes exist as monomeric aminoalkylsiliconates:

$$(-O)_3SiCH_2CH_2CH_2NH_2$$

The unique solubility of amino-organofunctional silanes is not limited to those with nitrogen on the third carbon atom from silicon. A series of ethylenediamine(En)-functional silanes, was prepared where the first nitrogen was 1 to 8 atoms removed from silicon. Stability of 10% aqueous solutions was observed on distilled samples of these materials (Table 3.4).

$$(MeO)_3Si-R-NHCH_2CH_2NH_2$$

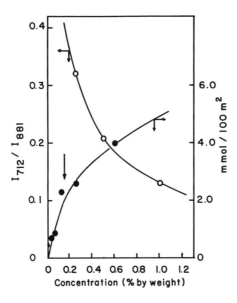

**Figure 3.6.** Amount of adsorbed $^{14}$C-labeled $\gamma$-APS from aqueous solution at various concentrations (from Ref. 4) and the relative Raman intensity $I_{712}/I_{881}$, read from Figure 3.3.

Initial solution may involve silanols stabilized through hydrogen bonding, but on aging, concentrated solutions equilibrate to highly cyclized siloxanes that retain water solubility if the total substituent is hydrophilic enough.

Reaction products of $(MeO)_3Si(CH_2)_3En$ with alkyl halides are monohydrohalides of substituted diamines and show some very unique solubility characteristics. The amine hydrohalide is strongly hydrophilic,

**Table 3.4.** Solubilities of Aminofunctional Silanes in Water

| $(MeO)_3Si-R-En$ Structure $-R-$ | Initial solution | Stability of 10% solution |
|---|---|---|
| $-CH_2-$ | Clear | $a$ |
| $-(CH_2)_3-$ | Clear | $a$ |
| $-(CH_2)_5-$ | Clear | $a$ |
| $-(CH_2)_6-$ | Sl. haze | 40 days |
| $-CH_2CH_2C_6H_4CH_2-$ | Haze | 17 days |
| $-(CH_2)_3OCH_2CH(OH)CH_2-$ | Clear | $a$ |

$^a$ Stable more than 3 months.

while the R group and *unhydrolyzed* $(MeO)_3SiCH_2CH_2CH_2—$ group are hydrophobic. Solubility characteristics of a series of compounds of Structure I were observed where R varied from methyl to very large hydrocarbon groups.[18] In each case, the crude product was tested for water solubility by observing a 5% dispersion in water. A portion of each product was prehydrolyzed with 1.5 mol water in methanol and observed again as a 5% dispersion in water (Table 3.5). A general pattern is evident: When R is less than three carbon atoms, the compound is soluble in water either directly or after prehydrolysis. Compounds with $R = C_3-C_{10}$ are insoluble

$$(MeO)_3SiCH_2CH_2CH_2NHCH_2CH_2NHR\cdot HX \qquad (I)$$

initially, but become soluble when prehydrolyzed. Compounds with $R = C_{12}-C_{18}$ are soluble directly in water, but are insoluble if prehydrolyzed in concentrated homogeneous solution; compounds with $R > C_{20}$ are insoluble in water under all circumstances.

This behavior is readily understood when it is recognized that the unhydrolyzed trimethoxysilyl group is somewhat hydrophobic, but that it becomes hydrophilic when hydrolyzed. The chemical nature of the hydrolyzed species is not known, but in water or alcohol solution the intermediate silanols appear to form strong hydrogen bonds with solvent and with amine groups (intramolecular cyclics or intermolecular oligomers) that stabilize the solution against gelation.

**Table 3.5.** Aqueous Solubility of
$(CH_3O)_3SiCH_2CH_2CH_2NHCH_2CH_2NHR\cdot HCl$

| R-Group structure | 5% Dispersion in water | |
|---|---|---|
| | Direct | Prehydrolyzed |
| $—C_2H_5$ | Clear | Clear |
| $—CH_2CH=CH_2$ | Hazy | Clear |
| $—CH_2C_6H_5$ | Cloudy | Clear |
| $—CH_2C_6H_4—CH=CH_2$ | Cloudy | Clear |
| $—CH_2C_6H_4—O—C_6H_5$ | Cloudy | Clear |
| $—n\text{-}C_4H_{13}$ | Cloudy | Clear |
| $—n\text{-}C_{10}H_{21}$ | Cloudy | Clear |
| $—n\text{-}C_{11}H_{23}$ | Cloudy | Hazy |
| $—Coco^a$ | Clear | Cloudy |
| $—Tallow^b$ | Clear | Cloudy |
| $—CH_3C_6H_4—C_{11}H_{23}$ | Hazy | Cloudy |
| $—Arichidyl\text{-}behenyl^c$ | Insoluble | Insoluble |

$^a$ R-distribution: 2-$C_{10}$, 53-$C_{12}$, 24-$C_{14}$, 11-$C_{16}$, and 10-$C_{18}$.
$^b$ R-distribution: 4-$C_{14}$, 29-$C_{16}$, and 67-$C_{18}$.
$^c$ R-distribution: 10-$C_{14}$, 80-$C_{20}$, and 10-$C_{22}$.

Compounds with $R = C_3-C_{10}$ form relatively stable emulstions of unhydrolyzed monomers with amine-hydrochloride groups at the water interface and trimethoxysilane groups oriented toward the center of the micelle. When the R group is $C_{12}-C_{18}$, molecular orientation is reversed in the micelle with the long alkyl groups in the center and methoxysilanes at the water interface where they are rapidly hydrolyzed and stabilized as hydrogen-bonded silanol groups.

Amine- or amine-hydrochloride-functional silanes, after hydrolysis in concentrated alcoholic solution, are very effective primers.[19] Films laid down from such primers are more flexible and more organic-compatible than comparable films derived from in situ hydrolyzed monomers or from dilute aqueous solutions of the same silane. Prehydrolysis in a concentrated solution apparently causes internal cyclization of siloxane structure to give a less highly cross-linked cured film. The internal cyclized structures, however, are less effective as coupling agents when applied from very dilute solutions to glass fibers or particulate mineral fillers. The difference may be that primers are deposited in much thicker films than are used in most coupling applications. Primer performance is thus dependent on mechanical properties of films, as well as on the chemistry of surface functional groups. Coupling agents applied to glass fibers are from such dilute solutions (e.g., about 0.1%) that film properties are of little significance. Thw function of such treatment is strictly for surface modification at the interface. An oriented monolayer is more probable from application of the open hydrolyzed silane than from a cyclic cage structure of the prehydrolyzed coupling agent.

A convenient way to prepare primers from amino-organofunctional silanes is to add about 5 parts of water to 100 parts of a 50% solution of silane in methanol and allow the mixture to age a few hours. There is an initial increase in viscosity as the silane hydrolyzes and condenses, but then a decrease in viscosity as the product equilibrates to its final stable cage form. The equilibrated primer solutions are stable indefinitely and may be diluted further with alcohol, toluene, etc., for application as primers.

The substituted aminofunctional silane (Structure I) (where R is a vinylbenzyl group) formed cloudy dispersions in water. After centrifuging, oily droplets were recovered and shown to contain unhydrolyzed methoxysilane groups. When the same compound was prehydrolyzed as a 40% solution in methanol, the product was completely soluble in water. When Structure (I) was partially prehydrolyzed by adding 1% water to a 40% solution of silane in methanol, the product gave a very smooth, stable dispersion in water. The small amount of soluble, hydrolyzed silane acted as emulsifier for the bulk of unhydrolyzed silane to improve stability of aqueous dispersions.

The reaction product of octadecyltrichlorosilane and 3 mol of an alkanolamine is soluble in water to form clear dilute solutions that are stable for several hours.[20] Because octadecylsilanetriol is not soluble in water and is not solubilized by adding alkanolamines, it appears that the beta-substituted alkoxy groups on silicon have moderate stability in neutral water and are hydrophilic enough to hold the compound in solution:

$$C_{18}H_{37}SiCl_3 + 3HOCH_2CH_2NR_2 \rightarrow C_{18}H_{37}Si(OCH_2CH_2NR_2HCl)_3$$

### 3.3.2. Quaternary Ammonium-Functional Silanes

Quaternary ammonium-functional silanes are formed readily by the reaction of chloroalkyltrimethoxysilanes and tertiary amines. Such products are generally soluble in water to form stable solutions:

$$(MeO)_3SiCH_2CH_2CH_2Cl + CH_2{=}C(CH_3)COOCH_2CH_2NMe_2 \rightarrow$$

$$CH_2{=}C(CH_3)COOCH_2CH_2N^{\oplus}(Me)_2CH_2CH_2CH_2Si(OMe)_3Cl^- \quad (A)$$

$$(MeO)_3SiCH_2CH_2CH_2Cl + C_{18}H_{37}NMe_2 \rightarrow$$

$$C_{18}H_{37}N^{\oplus}(Me)_2CH_2CH_2CH_2Si(OMe)_3Cl^- \quad (B)$$

$$(MeO)_3SiCH_2CH_2CH_2Cl + HOCH_2CH_2NMe_2 \rightarrow$$

$$HOCH_2CH_2^{\oplus}(Me)_2CH_2CH_2CH_2Si(OMe)_3Cl^- \quad (C)$$

Compound (A) forms stable solutions in water and is used to treat fiberglass reinforcements for unsaturated polyesters. It imparts antistatic properties to glass[21] in addition to coupling to resin.

Compound (B) is stable in neutral or acidic aqueous solutions, but precipitates from alkaline solutions. It is very substantive on any solid surface with an isoelectric point less than 7 and makes the surface strongly hydrophobic. Its major application is related to its biocidal activity.[22]

Compound (C) forms stable solutions in water at all pH's and provides a hydrophilic cationic surface treatment. In acid solutions at pH less than 4, it is a powerful solubilizer for aqueous silicates.

## 3.4. Anionic-Organofunctional Silanes

Many anionic-functional silanes are prepared in water and have such high solubility in water that they are not isolated, but are used directly as aqueous solutions. Undesired counter ions may be moved with cation-exchange resins, but this is generally not necessary.

Carboxyphenylsiliconates, when freshly precipitated by acidifying aqueous solutions of the sodium salts, formed stable solutions in 50/50 acetone/water or in alcohols, but could not be redissolved in water. Ion exchange of the sodium salt, however, gave dilute solutions in water that were stable for several days at pH 3–4.[23]

Reaction products of primary and secondary aminofunctional silanes with equimolar (per amine group) cyclic anhydrides are acid-functional amides with good solubility in neutral or alkaline solutions. The product of $N$-(2-aminoethyl)-3 aminopropyltrimethoxysilnae with 2 mol of maleic anhydride is reported to be a good coupling agent for unsaturated polyesters and many thermoplastic polymers.[24] The same diamine-functional silane with a slight excess of isophthalic acid is a convenient acid-functional silane for "ionomer bonding" of polymers to silane-treated fillers (cf. Section 2.4.6).

Phosphorus-functional silanes are prepared by warming chloralkyl-silanes with methyl esters of phosphorus in the presence of amine catalysts.[25] Saponification of remaining methyl ester group gives anionic phosphorus-functional silanes:

$$(MeO)_3SiCH_2CH_2CH_2Cl + MeP(O)(OMe)_2$$

$$\xrightarrow{R_3N} (MeO)_3SiCH_2CH_2CH_2OPO(OMe) \qquad (I)$$

$$(I) + NaOH \xrightarrow{H_2O} O_{1.5}SiCH_2CH_2CH_2OP(O)(CH_3)ONa$$

These materials are very soluble in water over the entire pH range and are effective stabilizers for aqueous silicates.

## 3.5. Zwitterion-Organofunctional Silanes

Reaction products of 1 mol of cyclic carboxylic anhydride with diamine-functional silanes are zwitterions and are soluble in either acid or alkaline solutions, but may be insoluble in neutral water. A vinylbenzyl cationic silane with the structure

$$(CH_3O)SiCH_2CH_2CH_2NHCH_2CH_2NHCH_2C_6H_4CH{=}CH{\cdot}HCl \qquad (I)$$

is a good adhesion promoter for many thermoplastic polymers, but it cannot be added to many anionic emulsions without causing them to coagulate. The zwitterion reaction product of (I) with succinic anhydride is also a good adhesion promoter and has good compatibility with common anionic latexes.[26]

Very effective chelating agents related to ethylenedianime tetraacetic acid are prepared from a diamine-functional silane and 3 mol of chloroacetic acid in the presence of 6 mol of sodium hydroxide. These materials, bonded to a siliceous surface such as silica gel, are very effective ion exchange media,[27] and in water are good stabilizers for sodium silicate solutions.

### 3.6. Stabilization of Aqueous Silicate Solutions

Sodium silicates are manufactured by fusing silica and alkali in various ratios of $SiO_2:Na_2O$. Theoretically, any ratio of silica and alkali may be used, but soluble silicates do not exceed a ratio of about 4:1.

Aqueous solutions of sodium silicate are more or less polymeric, depending on the ratio of silica to alkali. Similarly, partial neutralization of the alkali in sodium silicate solutions will increase the molecular weight of the polysilicate until the solution finally gels.

Rapid acidification of sodium silicate solutions may produce soluble polysilicic acids that have limited stability. Monomeric silicic acid may be obtained by mild acid hydrolysis of ethyl silicate. Silicic acid condenses to polysilicic acids and eventually gels:

$$Si(OEt)_4 \xrightarrow{H_2O/H+} Si(OH)_4 + 4EtOH$$

Stability of silicic acid in water is a function of concentration and pH, as indicated[28] in Figure 3.7. Silanol groups have maximum stability at a pH of 2–3. Condensation rate to polysilicates is greatest at a pH of about 8. At high pH, the silicates are permanently soluble as silicate ions.

Alkaline silicate solutions will equilibrate with soluble organic siliconates. Condensation and gelation of such mixtures is not greatly different from condensation and gelation of silicates alone, because the siliconates are also polyfunctional and cross-link through silanol condensation:

Silicate     Siliconate     Mixed silicate/siliconate

$$
\begin{array}{ccc}
\overset{\displaystyle O}{\underset{\displaystyle O}{\overset{|}{\underset{|}{O-Si-O}}}} & + & \overset{\displaystyle O}{\underset{\displaystyle R}{\overset{|}{\underset{|}{O-Si-O}}}} \quad \rightleftarrows \quad \overset{\displaystyle O}{\underset{\displaystyle O}{\overset{|}{\underset{|}{O-Si-O}}}} \overset{\displaystyle O}{\underset{\displaystyle R}{\overset{|}{\underset{|}{-Si-O-}}}}
\end{array}
$$

When the organofunctional group on silicon is strongly cationic or anionic, the mixture may have complete stability under conditions where silicates alone would gel. Silicates are stabilized by certain stable ionic

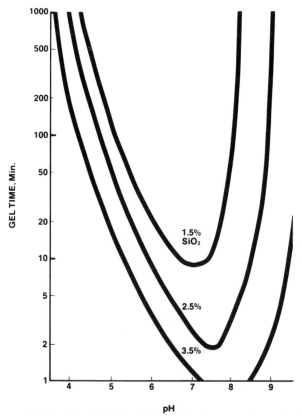

**Figure 3.7.** Stability of silicic acid in water vs. concentration and pH.

surface charges in the growing polysilicate structure:

$$
\text{Polysilicate}\left\{
\begin{array}{c}
| \\
-\text{O}-\text{Si}-\text{R}^{\oplus} \\
| \\
\text{O} \\
| \\
-\text{O}-\text{Si}-\text{R}^{\oplus} \\
|
\end{array}
\right.
$$

$$
\text{Polysilicate}\left\{
\begin{array}{c}
| \\
-\text{O}-\text{Si}-\text{R}^{\ominus} \\
| \\
\text{O} \\
| \\
-\text{O}-\text{Si}-\text{R}^{\ominus} \\
|
\end{array}
\right.
$$

As might be expected, cationic surface charges are most effective at low pH. Anionic structures are most effective at higher pH, but have some stabilizing effect over the entire pH range.

Table 3.6. Stability of 1 Molal Siliconate/Silicate Mixture at pH 8

| | Time to gel | |
|---|---|---|
| Functional group on siliconate | 1 : 5 mol ratio | 1 : 10 mol ratio |
| None (silicate alone) | 2 sec | |
| —CH$_3$ | 35 sec | |
| —C$_6$H$_5$ | 15 sec | |
| —CH$_2$CH$_2$SH | 40 sec | |
| —CH$_2$CH$_2$CH | 2 min | |
| —CH$_2$CH$_2$OH | 1 hr | |
| —CH$_2$CH$_2$CH$_2$NHCH$_2$CH$_2$NH$_2$ | 10 sec | |
| —CH$_2$CH$_2$CH$_2$COONa | $a$ | 4 hr |
| —CH$_2$CH$_2$CH$_2$OPO(Me)ONa | $a$ | 2 days |

$a$ No gel in 7 days.

A simple test was developed to compare organofunctional silanes as stabilizers for silicates. One molal solutions of organofunctional siliconates were mixed with various ratios of 1 $m$ sodium silicate (weight ratio SiO$_2$ : Na$_2$O = 3.22) and allowed to age for at least 1 hr. The mixtures were then neutralized with 10% aqueous HCl to pH 8 and observed for time of gelation at room temperature. Compositions that did not gel in 7 days generally were stable indefinitely (no gel in 6 months). Siliconates that imparted stability at 1 : 5 ratios with silicate were also tested at higher silicate ratios. Initial tests showed that hydrophilic organofunctional silanes only delayed gelation of silicates. Certain anionic-functional silanes imparted complete stability at a 1 : 5 mol ratio in 1 $m$ polysilicate (6% SiO$_2$) (Table 3.6).

Equilibration of siliconate with aqueous sodium silicate is rapid at room temperature, as indicated by stability of the mixture at pH 8. Portions of a 5 : 1 mol mixture of 1 $m$ sodium silicate and phosphonate-functional siliconate of Table 3.6 were neutralized to pH 8 at various time intervals

Table 3.7. Stability of 1 Molal Siliconate/Silicate
Mixture vs. Time at pH 8

| Time for equilibration of alkaline mixture | Time to gel at pH 8 |
|---|---|
| 10 sec | 30 sec |
| 1 min | 70 sec |
| 5 min | 20 min |
| 15 min | $a$ |
| 45 min | $a$ |

$a$ No gel in 7 days.

**Table 3.8.** Stability of 1 Molal Anionic Siliconate/Silicate Mixtures

| Functional groups on siliconate | Time to gel at mol ratios indicated by pH 8 | | | |
|---|---|---|---|---|
| | 1:5 | 1:10 | 1:15 | 1:20 |
| —CH$_2$CH$_2$COONa | $a$ | 4 min | 2 min | |
| —CH$_2$CH$_2$SCH$_2$COONa | $a$ | $a$ | 3 min | 90 sec |
| —CH$_2$CH$_2$SCH$_2$CH$_2$COONa | 2 days | 20 sec | | |
| —CH$_2$CH$_2$CH$_2$N(CH$_2$COONa)$_2$ | $a$ | 45 sec | | |
| —CH$_2$CH$_2$CH$_2$NHCH$_2$CH$_2$N(CH$_2$CH$_2$COONa)$_2$ | $a$ | $a$ | 1 min | 10 sec |
| —CH$_2$CH$_2$CH$_2$N(CH$_2$COONa)CH$_2$CH$_2$N(CH$_2$COONa)$_2$ | $a$ | $a$ | $a$ | 90 sec |
| —CH$_2$CH$_2$CH$_2$OCH$_2$CH(OH)CH$_2$SO$_3$Na | $a$ | $a$ | $a$ | $a$ |

$a$ No gel in 7 days.

**Table 3.9.** Stability of Molal Modified Polysilicic Acids (1:5 mol ratio siliconate/polysilicate).

| Functional group on siliconate | Time to gel at indicated pH | | | | | | | |
|---|---|---|---|---|---|---|---|---|
| | 2 | 3 | 4 | 5 | 6 | 7 | 8 | 9 |
| None (Polysilicate alone) | 3 day | 4 hr | 25 min | 10 min | 45 sec | 1 sec | 1 sec | 1 sec |
| —$CH_2CH_2CH_2NHCH_2CH_2CH_2NH_2$ | a | a | 3 hr | 40 min | 1 min | 1 sec | 1 sec | 1 sec |
| —$CH_2CH_2CH_2N^{\oplus}(Me_2)_2CH_2CH_2OH \cdot Cl^-$ | a | a | 4 day | 4 hr | 1 hr | 2 day | a | a |
| —$CH_2CH_2CH_2COONa$ | 5 hr | 30 min | 15 min | 23 hr | 20 hr | a | a | a |
| —$CH_2CH_2CH_2OPO(Me)ONa$ | a | 6 day | 2 day | 4 hr | 1 day | a | a | a |
| —$CH_2CH_2CH_2OPO(Me)ONa$ (at 2:5 mol ratio) | a | a | a | a | a | a | a | a |

[a] Still stable after 7 weeks.

after mixing and observed for stability (Table 3.7). An equilibration time of 15 min at room temperature was sufficient to give a stable product.

Some anionic-functional siliconates were tested at pH 8 with the same sodium silicate at higher silicate ratios. As much as 15 mol of silicate could be stabilized by 1 mol of some polyanionic siliconates (Table 3.8).

No definite pattern of silicate stabilization can be discerned. Just a slight difference in structure of carboxylicfunctional siliconate makes a great deal of difference in stabilizing capacity. Polyfunctional carboxylic acids are generally better than monocarboxylic acids. Some zwitterion-(amino acid) functional silanes are very effective. The strongly anionic sulfonate siliconate and a phosphonate-functional siliconate are exceptionally good silicate stabilizers.[29]

Stability of 1:5 mol ratios of 1 molal siliconate/silicate were compared over the pH range of 2 to 9 with several of the organofunctional siliconates. The equilibrated mixtures were acidified quickly with concentrated HCl to a pH of 2. Portions of this solution were adjusted back to higher pH with 10% aqueous sodium acetate solution (Table 3.9).

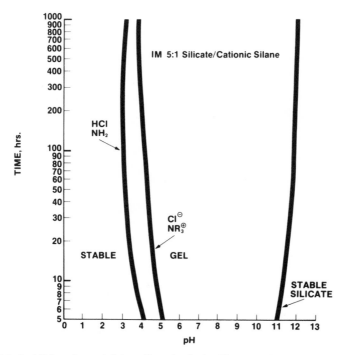

**Figure 3.8.** Stabilities of 1 molal 5:1 silicate/cationic siliconate mixture over the complete pH range.

Stabilities of 5:1 mixtures of 1 $m$ sodium silicate and cationic silanes are shown in Figure 3.8. They show improved stability only in the acid range. Quaternary ammonium-functional silanes are somewhat more effective than amine-hydrochloride-functional silanes.

Anionic-functional siliconates are effective stabilizers for alkali silicate corrosion inhibitors in alcohol compositions used as coolants in cooling systems of internal combustion engines.[30] It is expected that they will also be effective in preventing precipitation of silica from boiler water and steam, and as stabilizers for silica sols.

## References

1. E. R. Pohl. SPI, 38th Ann. Tech. Conf. Reinf. Plast. 4-B (1983).
2. E. R. Pohl and F. O. Osterholz. In *Molecular Characterization of Composite Interfaces*, H. Ishida and G. Kumar, Eds., p. 162. Plenum, New York (1985).
3. H. Ishida and J. L. Koenig. *Appl. Spectroscopy* 32, 469 (1978).
4. E. P. Plueddemann. SPI, 24th Ann. Tech. Conf. Reinf. Plast. 19-A (1969).
5. E. R. Pohl and F. O. Osterholz. In *Silanes, Surfaces and Interfaces*, Vol. 1, D. E. Leyden, Ed., pp. 481–500, Gordon & Breach, Amsterdam (1986).
6. C. W. Lentz. *Inorg. Chem.* 3, 574 (1964).
7. H. Ishida and J. D. Miller. *Macromolecules* 11, 1659 (1984).
8. K. A. Andrianov and B. A. Izmaylov. *J. Organomet. Chem.* 8, 435 (1967).
9. M. M. Sprung and F. O. Guenther. *J. Polymer Sci.* 28, 17 (1958).
10. J. F. Brown, Jr., and L. H. Vogt. *J. Am. Chem. Soc.* 84, 4313 (1965).
11. J. F. Brown, Jr. *J. Am. Chem. Soc.* 84, 4317 (1965).
12. H. Ishida, S. Vaviroj, S. K. Tripathy, J. J. Fitzgerald, and J. L. Koenig. SPI, 36th Ann. Tech. Conf. Reinf. Plast. 2-C (1981).
13. B. W. Pearce, K. G. Mayhan, and J. F. Montle. *Polymer* 14 (Sept), 420 (1973).
14. E. P. Plueddemann. SPI, 36th Ann. Tech. Conf. Reinf. Plast. 4-C (1984).
15. C. L. Frye, G. E. Vogel, and J. A. Hall. *J. Am. Chem. Soc.* 83, 996 (1961).
16. C. H. Chiang, H. Ishida, and J. Koenig. *J. Colloid Interface Sci.* 74(2), 396 (1980).
17. O. K. Johannson, F. O. Stark, G. E. Vogel, and R. M. Fleishmann. *J. Compos. Mater.* 1, 278 (1970).
18. E. P. Plueddemann. *Silyated Surfaces*, D. E. Leyden & W. Collins, Eds., p. 40, Gordon & Breach, New York (1980).
19. B. M. Vanderbilt and R. E. Clayton. U.S. Patent 3,350,345 (to Esso) (1967).
20. C. L. Frye. U.S. Patent 2,14,572 (to Dow Corning) (1957).
21. E. P. Plueddemann. U.S. Patent 3,734,763 (to Dow Corning) (1973).
22. A. J. Isquith, E. A. Abbott, and P. A. Walters. *Appl. Microbiol.* 24(6), 59 (1973).
23. R. Wong and J. C. Hood. AFML-TR-65-316 Summary Technical Report, July (1965).
24. E. P. Plueddemann. U.S. Patent 4,718,944 (to Dow Corning) (1988).
25. E. P. Plueddemann. U.S. Patent 4,093,641 (to Dow Corning) (1978).
26. E. P. Plueddemann. U.S. Patent 3,956,353 (to Dow Corning) (1976).
28. R. C. Merrill and R. W. Spencer. *J. Phys. Colloid Chem.* 54, 506 (1950).
29. E. P. Plueddemann. U.S. Patent 4,370,255 (to Dow Corning) (1983); and U.S. Patent 4,503,242 (to Dow Corning) (1985).
30. A. N. Pines and E. A. Zientek. U.S. Patent 3,198,820 (to U.C.C.) (1965).

# 4 | Surface Chemistry of Silanes at the Interface

## 4.1. Introduction

Practical experience with the silane coupling agents in reinforced composites indicates that only a very small proportion of silane at the interface is sufficient to provide marked improvements in composite properties. The small amounts involved, and the fact that silanes are buried in the matrix resin during fabrication, makes it very difficult to study the nature of the coupling action by ordinary chemical analysis. Prototype chemical reactions have been demonstrated between coupling agents and matrix resins, but it is possible that reactions may be quite different when the silane is aligned in a very thin film on a solid substrate rather than dissolved in the matrix resin.

A chemical bonding theory has been the favorite working hypothesis in explaining the action of silane coupling agents. The simplest view pictures a mineral substrate with a film of coupling agent as a monolayer—with each molecule chemisorbed to the surface through silanol reactions—while another functional group of each molecular remains available for reaction with the matrix resin. In practice, a monolayer coverage is rarely observed, but the concept is useful in predicting performance of coupling agents in composites.

Any modification in the interface between resin and reinforcement has far-reaching effects on the resin adjacent to the treated mineral extending into a broader interphase region. Solubility effects, or polar effects, may alter the orientation of polymer segments in the interphase region. Catalytic effects of the surface may contribute higher or lower degrees of cross-linking in the interphase region.

Some of the surface chemistry of silane coupling agents may be deduced from indirect evidence of composite properties. Performance studies of this nature have brought composite structures to a high degree of sophistication, but their production is still much of an art and reproducibility leaves a lot to be desired. A beginning has been made in studying the surface chemistry of composite interfaces by advanced analytical techniques, but there is still much to learn about the interface and interphase regions of composites.

## 4.2. Instrumental Techniques

The surface of a specimen may be defined as the region from which information is received in a particular measurement. Under this definition, surfaces may be examined by exposing the surface to a variety of controlled radiations and examining the emitted radiation by instrumental techniques. Optical microscopy is of limited value in examining the interface region, since resolution is limited by the wavelength of visible light to about 10,000 Å. By using an electron beam for imaging, it is possible to obtain resolution down to about 100 Å. Scanning electron microscopy (SEM) is therefore used to study details of fracture surfaces, but cannot resolve to monomolecular coverage.

High-energy electron beams penetrate the solid and cause emission of X rays that are analyzed as X-ray fluorescence (XRF). Elements within the solid may be identified by XRF, but penetration is too deep to be of much value in study of the interface.

Along with X-ray emission from electron bombardment, the energy may also be carried away by an electron characteristic of the de-excitation process recognized as an Auger electron. Auger electrons may be controlled to escape depths as shallow as 3 to 6 Å in some materials, and so are more useful than X rays for surface analysis (AES). The Auger line characteristic of a given element may have features that differ with different chemical bonding of the element.

Depth profiling of a surface region is possible while monitoring Auger signals of AES. This is done by sputtering away an area of the surface under constant bombardment by a beam of inert gas ions. Sputtering rates of the order of a monolayer per second are attainable.

Electron beams are readily focused and X-ray beams are not, but X rays offer some advantage as primary excitation beams. They provide a more penetrating radiation than electrons, and thus are less disruptive to fragile surfaces. X-ray beams are particularly valuable for study of insulators, because charging problems are much less severe than with electron beams. They also produce one-electron excitations (in addition to Auger emission) that may contain more readily accessible chemical information. Detection of this single photoelectron is the basis of electron spectroscopy for chemical analysis (ESCA), which is also known as X-ray photoelectron spectroscopy (XPS). ESCA provides a direct measure of binding energies of atomic levels and provides simpler line shapes than does AES.

In ion-scattering spectroscopy (ISS), a focused beam of inert gas ions is impinged on the surface to be studied. Ions scattered from the primary beam have energies and scattering angles resulting from target collisions at the surface. Peaks in the ISS spectra can be correlated with particular masses of the target atoms, and thus provide chemical analysis data.

Secondary ion mass spectroscopy (SIMS) also uses a focused beam of incident ions, but monitors the sputtered target ions with a mass spectrometer. Unit mass resolution is good, and even isotopic components are evident in spectra.

Several of these techniques are being incorporated in the same ultra-high vacuum chamber to provide versatile hybrid analysis for analytical studies. Combined ISS/SIMS analysis can determine surface composition, spatial distribution, concentration variation with depth, and, to some extent, bonding of species on a sample surface.

The most sensitive method of studying trace amounts of coupling agent on a surface is to use radioactive silanes tagged with $^{14}C$ in the organofunctional group. Surface coverage of a fraction of a monolayer has been measured by this technique.

All the above instrumental techniques are limited in their capability to observe and identify chemical reactions at the interface between coupling agent and mineral, or between coupling agent and matrix resin. Infrared (IR) spectroscopy is uniquely suited for studies of chemical reactions, because absorbance peaks are characteristic of chemical bonds rather than of atoms. Limited sensitivity of conventional IR is such that high-surface-area minerals must be used to get useful data. IR signals from plane surfaces may be multiplied by using internal or external multiple reflections.

Some materials such as KRS-5, Ge, and Si are transparent to IR radiation without a major loss in the energy level. When a sample is in contact with these plates, the IR radiation slightly penetrates the sample and yields information about its surface. The penetration depth is a function of the refractive index difference between the substrate and the sample, the angle of the incident radiation, and the frequency of the radiation. Because this technique utilizes totally reflected IR beams, it is called attenuated total (internal) reflection spectroscopy.

If highly reflective surfaces such as well-polished metal surfaces are used as substrates, external reflection spectroscopy can be used to study the surface species. Only the incident light with parallel polarization has a sufficient amplitude at the metal surface to interact with the surface species. High sensitivity and ability to obtain information on the orientation of the surface species are of particular interest with this technique, which is called reflection absorption spectroscopy.

Probably the most powerful tool for observing chemical structures and transformations at the interface is Fourier transform infrared spectroscopy (FT-IR). When a Michelson interferometer is used instead of an ordinary IR setup, the detected IR radiation forms an interferogram which is a function of time rather than frequency. This time-domain interferogram can be converted into the frequency-domain infrared spectrum through the mathematical operation called Fourier transform. By combining the spectrophotometer with a computer, it is possible to store multiple scans of the subject and let the computer subtract background signals, thus getting thousandfold increases in useful information.

When a microphone is used as a detector instead of an ordinary IR detector, photoacoustic spectrum in the IR region can be obtained. This technique may be suited for surface studies of the materials that are very opaque in the IR.

A relatively new technique, known as inelastic electron tunneling spectroscopy (IETS), has been used to study the adsorption of organic compounds and silane coupling agents on metals. A metal–insulator–metal junction is formed, and the current at different voltages is measured at 4.2 K. Aluminum is generally used as the substrate, although other metals may be used. Lead is the preferred counter electode because it becomes superconducting at this temperature. The polymer or silane insulator should be very thin, and may be as little as a fraction of a monolayer. Instrumentation for this technique is much less costly than that of the various high-vacuum radiation techniques used to study surfaces.

All the above instrumental techniques have been used to study silane coupling agents at the interface. As these techniques are refined, it should be possible to observe structures and reactions that have been inferred in the past from indirect evidence of composite performance. It should then be possible to design total composites with consistently good performance.

## 4.3. Identifying Silanes at the Interface

Early chrome finishes (Volan®) at the interface were readily identified by the blue-green color imparted to clear resin laminates. Chromium could also be leached from glass and identified by color reactions. Until 1960, the only silanes used to compete with chrome finishes were vinyl silanes and aminofunctional silanes. A simple acid–base indicator showed whether an aminofunctional silane was on the glass. If the glass did not have a chrome finish or an aminofunctional silane, it was assumed to have a vinylsilane finish.

The larger number of commercial silane coupling agents available today (Table 2.2) cannot be differentiated or identified so easily. A simple test with dilute permanganate solution shows significant differences among the silanes of Table 2.2. Pieces of heat-cleaned fiberglass cloth were dipped into 0.25% aqueous solutions of silanes and dried. A drop of 0.2 $N$ KMnO$_4$ solution placed on each was observed until the purple permanganate color was reduced to the brown color of manganese dioxide (Table 4.1). Mercapto- and amino-organofunctional silanes discolor permanganate rather rapidly, the styryl-functional amine being the most rapid. Methacrylate- and vinyl-silanes discolor permanganate more slowly. Chloropropyl- and epoxyfunctional silanes do not discolor permanganate. A methacrylate–chromium complex (Volan®) is somewhat more reactive than a methacrylate silane in discoloring permanganate solutions, possibly because the chrome-finished surface is more hydrophilic.

Any analytical test for silanes on reinforcements may be complicated by other ingredients in commercial sizes and finishes. Fiberglass manufacturers and weavers generally consider the composition of their sizes and finishes to be proprietary and only disclose their products to be "compatible"

**Table 4.1.** Permanganate Test for Silanes
(0.25% silane on glass cloth)

| Functional group on silane | Time to discolor permanganate |
|---|---|
| None | a |
| Vinyl | 20 min |
| Chloropropyl | a |
| Epoxy | a |
| Methacrylate | 5 min |
| Primary amine | 3 min |
| Diamine | 2 min |
| Mercapto | 1 min |
| Cationic Styryl | 1 min |
| Volan® (chrome complex) | 2 min |

$^a$ Still purple after one hour.

with specific resins. Filler suppliers generally disclose the specific silanes used in commercially available treated fillers.

X-ray fluorescence as an analytical tool may be applied to the determination of elements of atomic number greater than 11. It is not particularly suited for determining trace amounts of elements in aqueous solution. By preconcentrating cations[1] and anions[2] on silica gels, silylated with 3-(2-aminoethylamine)propyltrimethoxysilane, Leyden and co-workers were able to determine heavy metal ions in parts per billion concentration in water. Metals could be determined directly on the silica gel substrate by X-ray fluorescence (see Figures 8.3 and 8.4), or they could be eluted with dilute acid and determined by atomic absorption spectrometry. The capacity for heavy metal cations was generally close to one atom of metal per two moles of silylating agent, suggesting rather complete utilization of the diamine in 2:1 ion–metal complexes.

Electron spectroscopy for chemical analysis (ESCA) was used by Nichols et al.[3] to identify certain elements such as sulfur, chlorine, or nitrogen in silane coupling agents on a glass surface. The ESCA signal also varies with chemical structure, making it possible to differentiate between a primary amine, a diamine, or an amide in nitrogen-functional silane finishes.

Finkler and Murray[4] used ESCA to study the silanization of $TiO_2$ electrodes with 3-(2-aminoethylamino)propyltrimethoxysilane. An essential monolayer of silane was stable on the electrode through light-induced oxidation reactions, but had little effect on semiconductor properties. These results indicate that a large proportion of unreacted Ti—OH groups remained on the silanized surface.

A similar silane layer on $SnO_2$ was reacted with various difunctional acid chlorides.[5] Assays by ESCA showed that oxalyl chloride binds exclusively as the monoamide, whereas glutaryl chloride bonds primarily as the bridged diamide. These results are interpreted as evidence for broad spacing of amine groups on the silanated oxide surface resulting from internal cyclization of nitrogen on the 3-carbon atom with silanol of the coupling agent.

Auger electron spectroscopy (AES) probes the first few monolayers of a surface. When used with reactive ion etching (RIE), in which argon ions sputter off surface atoms, AES may be used to give a depth profile of a surface layer. Cain and Sacher[6] used an AES technique with RIE to study aminofunctional and epoxyfunctional silane layers deposited on a hyperpure silicon wafer. Compared with an untreated surface, a surface treated with aqueous aminofunctional silane (F of Table 2.2) showed carbon and nitrogen in a dense layer more than 60 Å thick. The epoxyfunctional silane

(C of Table 2.2) gave a much looser layer that allowed electrons to detect the silicon substrate even while sampling the first few angstroms of a layer more than 60 Å thick. When the aminofunctional silane was deposited from 50/50 water/methanol, it also gave a layer that allowed substrate penetration. Profiles of carbon and nitrogen in a well-bonded aminosilane film showed almost mirror image relationsips. Nitrogen is highest at the outer edge and at the interface, while carbon is lowest in these regions (Figure

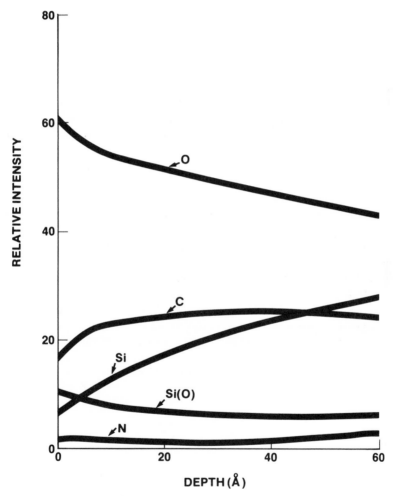

**Figure 4.1.** A thickness profile of a silane layer deposited from γ-aminopropyltriethoxysilane in water as observed by AES.

4.1). Because nitrogen and carbon are on the same molecule, such a profile indicates some preferential orientation of amino groups towards silica and air interfaces.

Wong[7] examined E-glass and S-glass fibers by Auger spectroscopy and observed that their surface compositions were much different from bulk glass compositions (Table 4.2). A typical E-glass surface is low in magnesium, boron, and calcium, but high in fluorine, silicon, and aluminum, while an S-glass surface is rich in magnesium and aluminum when compared to the bulk glass.

ESCA was used by Nichols and Hercules[8] to show the change in concentration of elements on E-glass surface when fiberglass is subjected to heat or acid treatment. Calcium diffused to the surface as a function of temperature from 26 to 750°C. Acid leaching removed both calcium and aluminum from the glass surface. E-glass that was silanized with either 3-(2–aminoethyl-amineo)propyltrimethoxysilane or 3-mercaptopropyltrimethoxysilane extracted heavy metal ions from solution. Benzenesulfonyl chloride gave a stable sulfonated surface with the amine-silanized glass, but reacted with only one of the available nitrogens.

Di Benedetto and Scola[9] described the use of combined ion scattering spectroscopy and secondary ion mass spectrometry to evaluate adhesive failures in S-glass polysulfone composites. By analyzing failure surfaces, it was possible to locate the locus of failure. With clean S-glass in polysulfone, shear failure occurred at $12.1 \times 10^6 \, N/m^2$ by cleavage in the polysulfone within 30 Å of the glass surface. In composites of S-glass coated with APS ($\gamma$-aminopropyltriethoxysilane) shear failure again occurred by cleavage in the polysulfone within 20 Å of the glass, but at $21.5 \times 10^6 \, N/m^2$. In both cases, high concentrations of alkali metals and fluorine were found in the polysulfone to a depth of 100 Å. Migration of impurities may have weakened the polymer adjacent to the glass. The silane coupling agent in some way increased the strength of adjacent polymer in spite of the presence of the

**Table 4.2.** Composition of Glass Fibers

|        | Atoms per hundred atoms of "glass" | | | | | | | |
|--------|------|------|------|------|------|------|------|-------|
| Glass  | B    | Ca   | O    | F    | Mg   | Al   | Si   | Na    |
| E-Glass bulk    | 4.5 | 6.4 | 61.8 | 0.32 | 2.3  | 5.8  | 18.8 | Trace |
| E-Glass surface | 3.4 | 1.9 | 61.2 | 1.4  | 0.6  | 7.0  | 24.3 | —     |
| S-Glass bulk    | —   | —   | 62.2 | —    | 4.9  | 9.4  | 21.4 | —     |
| S-Glass surface | —   | —   | 52.0 | —    | 17.2 | 12.7 | 18.1 | —     |
| A-Glass bulk    | —   | 3.6 | 60.9 | —    | 1.8  | 0.7  | 27.4 | 5.6   |

same impurities. It would be interesting to determine the effect of water immersion on the mode of failure in the same composites.

## 4.4. Thickness of Silane Layers

### 4.4.1. General

Rough estimates of the thickness of silane layers on mineral reinforcement surfaces may be made by measuring the amount of silane picked up (weight gain) per unit area of surface. Silane on finished glass may be estimated from total carbon analysis, or from ignition loss (weight loss).

Commercial glass fabric shows an ignition loss of about 0.1 to 0.2%, while commercial glass roving shows an ignition loss of 1 to 2% of the weight of glass. Silanes comprise most of the glass cloth finish, but only about 10% of the roving size. Assuming that 10 $\mu$ fibers have a surface area of about 0.1 m$^2$/g, and that a silane molecule RSiO$_{1.5}$ covers an area of about 50 Å$^2$, a monolayer of silane coupling agent would comprise about 0.01% of the weight of finished glass. Commercial glass, then, has an average coverage of 10 to 20 monolayers of silane coupling agent. It has been found empirically that increasing or decreasing the silane coverage significantly from this level causes a drop-off in composite performance.

Favis et al.[10] studied the deposition of silanes MPS and CSS (Table 2.2) on mica from dilute (0.025%) aqueous solutions. Both silanes were deposited in discrete steps with time as shown by carbon–hydrogen–nitrogen analyses. Each plateau corresponded to monolayer coverage and indicated close-packing and perpendicular orientation on the mica surface.

Belton and Joshi[11] used etching experiments on polimide-coated silicon wafers to determine the optimum thickness of the underlying APS coupling agent layers. The time required for removal of the polyamide film increased fourfold as the silane thickness increased from a monolayer to about 10 monolayer equivalents. Optimum thickness of silane primer layer for adhesion of polypropylene to aluminum was estimated by spreading measured volumes of dilute CSS primer over controlled areas of clean surface and assuming a specific gravity of 1.10 for the dried primer film.[12] Good adhesion of polypropylene (fused at 250°C) was obtained with primer thicknesses of 0.5 to 10 $\mu$. Adhesion dropped off with thinner or thicker primer layers.

These examples illustrate the difference in requirement for performance of silane coupling agents as adhesion promoters in finely divided reinforcements as contrasted to massive surface treatments. Finely divided reinforcements require only a chemical modification of the surface that can be accomplished by a few monolayers of silane. Primers on massive structures require a sufficiently thick hydrolyzed silane layer to contribute mechanical film properties and possibly solvent effects for the matrix resin. Similar differences were observed by Trivisonno *et al.* in comparing silanes as coupling agents in fiberglass-reinforced polyesters with the same silanes as primers on blocks of glass held together with adhesive.[13] Primer performance on blocks of glass was improved by modifying vinylsilanes with compatible polymers. The same polymer modification gave decreased performance as coupling agent in fiberglass composites.

### 4.4.2. Electron Microscopy

It is not possible to obtain direct magnification of glass surfaces with a transmission electron microscope. Glass fibers, either bare or treated with a coupling agent, were preshadowed with platinum and carbon. The glass was then removed in hydrofluoric acid and the replica examined in the electron microscope. Because the practical resolution for these specimens was between 40 and 50 Å, a perfect monolayer of coupling agent on glass could not be seen by this technique. Sterman and Bradley[14] showed that much of the silane applied (even at 0.1% average weight percent pickup) was found in capillary channels formed along contacts of glass filaments.

Vogel et al.[15] aspirated coupling agent solution rapidly from treated glass rather than air-drying after dip coating. They found no evidence of coupling-agent agglomeration, but observed fairly uniform coverage at levels up to 10 theoretical monolayers of coupling agent on glass. Leaching for 2 hr with boiling water removed 90% of the silane, leaving islands of coupling agent film on the surface. It was also noted that when 1 g of treated E-glass was boiled in 50 ml of distilled water, the pH of the water increased from 6.4 to 9.5 within 2 hr. These observations suggest strongly that the displacement of coupling agent from E-glass by boiling water occurs by attack on the glass substrate, with relatively large segments of coupling agent being removed by the mechanical action of the boiling water.

Very clear pictures of glass fibers in composites may be obtained by scanning electron microscopes.[16] Again, the magnification is not sufficient to show the coupling agent layer, but differences in the nature of failure at fracture surfaces are clearly seen. Figures 4.2 and 4.3 show SEM pictures of fracture surfaces of epoxy composites prepared from glass that had been treated with a chrome-complex and with a silane-based finish (CS-429). Fracture after 72 hr in boiling water showed massive debonding of resin from chrome-treated glass, while the silane finish allowed complete retention of resin bond to glass.

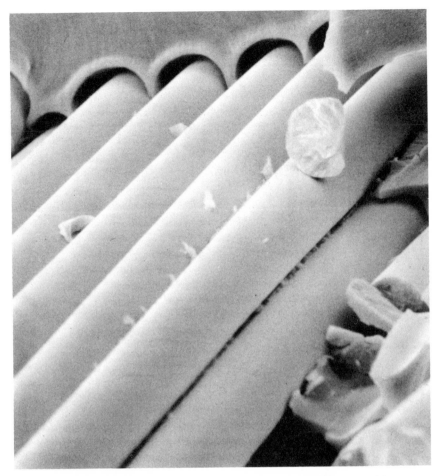

**Figure 4.2.** Fracture surfaces of epoxy composites after 72 hr in boiling water (chrome coupling agent).

### 4.4.3. Ellipsometry

The thickness of adsorbed organosilane films deposited on flat surfaces of glass and chromium by treatment with dilute aqueous solutions of silane coupling agents has been measured by optical ellipsometry techniques. Tutas et al.[17] observed that films deposited on glass from aminofunctional silane (APS) were the same thickness after a minute of contact as after an hour, while films adsorbed from aqueous vinylsilane coupling agent increased in thickness with time and had not leveled off after 100 min of contact with the solution. In all cases, films deposited from 0.1–5% aqueous

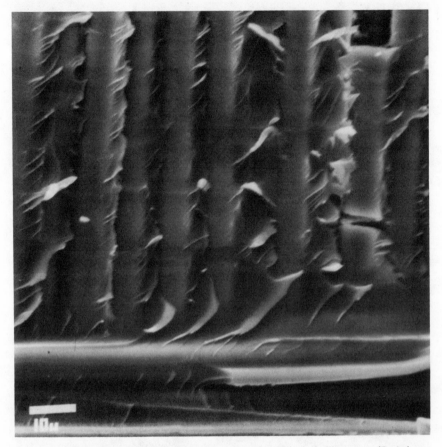

**Figure 4.3.** Fracture surfaces of epoxy composites after 72 hr in boiling water (Proprietary CS-429 coupling agent).

solutions were 50–200 Å thick, while monomolecular layers would be 5–10 Å thick. The aminofunctional silane adsorbed on chromium from 1% aqueous solution and the vinylsilane adsorbed on glass from 1% solution in methylethyl ketone were essentially of monomolecular thickness (8–10 Å).

Results of similar investigations by Bascom[18] of two chlorinated organofunctional silanes adsorbed on stainless steel gave values of $\gamma_c$ ranging from 44 to 47 dyn/cm, as would be expected of well-oriented adsorbed silanes. After rinsing the treated surface with acetone, ellipsometry showed that the films were definitely polymolecular, with a thickness of 233 Å for $ClCH_2CH_2CH_2Si(OCH_3)_3$ and 50–65 Å for $p\text{-}ClC_6H_4CH_2CH_2Si(OCH_3)_3$.

Observations of deposited silane films as measured by ellipsometry are in agreement with our understanding of the chemistry of organofunctional trialkoxysilanes in the presence of water (Chapter 3). Under mildly acidic conditions, neutral alkoxysilanes hydrolyze rapidly to form monomeric silanetriols and then condense slowly to polymeric organic siloxanols:

$$RSi(OCH_3)_3 \xrightarrow[\text{fast}]{H_2O} RSi(OH)_3 + 3CH_3OH$$

$$RSi(OH)_3 \xrightarrow[\text{slow}]{-H_2O} HO\left(\begin{matrix} R \\ | \\ Si-O \\ | \\ OH \end{matrix}\right)_n \begin{matrix} R \\ | \\ Si-OH \\ | \\ OH \end{matrix} + nH_2O$$

Amino-organofunctional trialkoxysilanes at their normal pH in water hydrolyze and condense very rapidly to reach an equilibrium composition in a short time. Very dilute aqueous solutions contain monomeric silanetriol, but more concentrated solutions are believed to be an equilibrium mixture of low-molecular-weight siloxanols with silanols stabilized by hydrogen bonding to amine groups (Section 3.3.1).

The thickness of siloxane film adsorbed on a mineral surface, therefore, depends on the nature of the organofunctional group on silicon, the availability of water, pH and age of the silane solution, topology of the surface, and presence or absence of specific catalysts. Ellipsometry measurements suggest that silane coupling agents, as commonly applied to mineral surfaces from water, do not deposit simple monomolecular films of silanols, but that the polymeric siloxane films do have a degree of orientation on the mineral sufficient to present a surface characteristic of the organofunctional group on silicon. A nearer approach to monomeric coverage is obtained when a mineral surface is treated with a dilute solution of an organofunctional silane in an anhydrous organic solvent.

### 4.4.4 Radioactive Tracers

By the use of [14]C-labeled silane coupling agents, it was possible to measure the adsorption of silane coupling agents on glass and silica down to fractions of monomolecular coverage. Schrader et al.[19] utilized radio-isotope techniques in combination with electron microscopy in an attempt to examine further the nature of a coupling agent film on glass. Of special interest was the question of the possible existence of a monolayer, which could not be detected with the electron microscope, in between the islands observed by Sterman and Bradley[14] and by Vogel et al.[15] The coupling agent chosen for investigation was $\gamma$-aminopropyltriethoxysilane (APS), with a $^{14}C$ label placed in the hydrocarbon chain alpha to the amino groups. The coupling agent was deposited on the surface of polished Pyrex blocks from benzene solution, and the surfaces of the blocks, exluding the edges, were then counted for radioactivity. The blocks were then extracted in water at room temperature, and the radioactivity on the blocks determined as a function of time of extraction. Following this, the blocks were placed in boiling water and the residual radioactivity was determined periodically.

It was found that the deposit of APS could be regarded as consisting of three fractions with respect to tenacity when subjected to extraction procedures. The major fraction, called fraction one, equalled as much as 98% of the total depending on the amount deposited and consisted of the hydrolyzate of the silane physically adsorbed to the surface. It could consist of as much as 270 monolayer equivalents and, although insoluble in benzene, was rapidly removed by the cold water rinse. Fraction two was a chemisorbed polymer of the coupling agent, consisting of about 10 monolayer equivalents. This fraction required about 3–4 hr of extraction in boiling water for essentially complete removal. Electron microscope pictures of this fraction taken in parallel experiments after 1 hr of boiling-water extraction showed the presence of islands. Upon complete removal of this second fraction, defined by the complete tapering off of the curve of radioactivity versus time obtained during the boiling-water extraction, Schrader et al.[19] found that the electron microscope now indicated a completely bare surface, all the islands having disappeared. However, a radioisotope count of this apparently bare surface indicated that the equivalent of a monolayer (or substantial fraction thereof) of the previously undetected coupling agent remained. This residue was called fraction three. It was apparently held more tenaciously to the surface than fraction two, and the authors speculated that this resulted from multiple bonding of the molecules (perhaps as dimers or trimers) to the glass surface.

Johannson et al.[20] utilized $^{14}C$-labeled silane coupling agents in an investigation of the nature of the adsorbed film of these agents on E-glass fibers. The coupling agents used were APS, MPS, and GPS. The authors obtained data on adsorption as a function of solution concentration by immersing heat-cleaned E-glass fibers in solutions of varying concentrations of each coupling agent. Each immersion, on a separate sample of fibers, took place for 1 min, after which the supernatant solution was removed by aspiration. In one experiment in which the immersion time was varied from 30 sec to 48 h, it was found that rapid adsorption took place within 30 sec, followed by a very slow further buildup of coupling agent. The authors reported about 10 monolayer equivalents (on the basis of their assumption of one molecule per 100 $\mathring{A}^2$) remaining on the surface, yielding a smooth appearance under the electron microscope. These results are essentially in accord with those of Sterman and Bradley[14] and of Schrader et al.[19] if allowance is made for differences in solution concentration and method of application (i.e., the method of Johannson et al.[20] minimizes the presence of fraction one).

An interesting difference is found in the surface coverage versus solution concentration among the three coupling agents investigated. MPS and GPS are neutral organofunctional silanes and showed an almost identical degree of coverage with concentration. Coverage increased at an increasing rate as concentration varied from 0 to 0.6%. Coverage by APS, on the other hand, showed a steep initial increase with concentration, but tapered off rapidly as concentration was increased above 0.1%.

The presence of coupling agents on the surface of glass in the form of a heterogeneous film raises the question as to which of the components of this film are responsible for its effectiveness in protecting the resin–glass interface from degradation by exposure to water. Schrader and Block[21] determined the efficacy of each of the fractions of the heterogeneous film by separately measuring the effectiveness of each in protecting glass–epoxy adhesive joints from the effects of moisture. Pyrex glass block surfaces containing selected components of the heterogeneous film, in varied amounts, were prepared by varying the nature of extraction with solvent following deposition of the coupling agents. The treated blocks were then crossed, in pairs, and cemented together with epoxy resin. Each glass–epoxy–glass adhesive joint was then immersed under a 50-lb load in hot water, and the time for failure of the joint ("joint life") was automatically recorded. It was found that joint lives of almost equal duration were obtained with or without the presence of fraction one (physically adsorbed hydrolyzate which is removed by cold water) in the heterogeneous coupling agent film. In fact, too large an excess of fraction one sometimes caused bond deterioration. Maximum joint lives were obtained with maximum amounts of fractions two and three (chemisorbed polymeric APS). With depletion of fraction two by extraction with boiling water before preparing the adhesive joint, there was a linear decrease in joint life (Figure 4.4). These tests were an evaluation of coupling agent layers as primers on bulk glass. As pointed out earlier (Section 4.3.1), the minimum silane coverage for bonding resin to finely divided reinforcements may be somewhat less.

Tests on removal of coupling agent layers from a glass surface by boiling water are not completely analogous to their performance as coupling agents in composites. As will be shown in Chapter 5, the water resistance

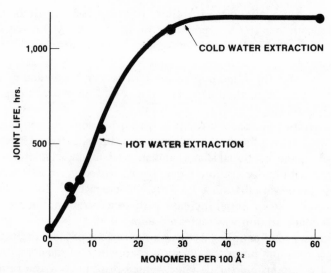

**Figure 4.4.** Joint life as a function of APS applied to surface by deposition–extraction method (Schrader and Block[21]).

of a coupling agent layer on glass is strongly dependent on the degree of cross-linking in the interface region. Reaction between the organofunctional group of the silane and reactive groups in the resin contributes a major proportion of cross-linkage at the interface.

## 4.5. Orientation of Layers

### 4.5.1. General

Some suggestions on orientation of organofunctional silanes on silica were mentioned earlier in Sections 4.2, 4.3, and 4.4. Auger electron spectroscopy identified a preferential orientation of amine groups of APS toward silica and air in multimolecular layers on silica.[6]

Orientation effects of silanes extend beyond the silane layer itself, as shown by orientation of organic liquid crystals on silane-treated surfaces. In this way, Kahn[22] obtained perpendicular orientation of liquid crystals on surfaces treated with 3-(trimethoxysilyl)propyldimethyloctadecyl ammonium chloride, and parallel alignment on surfaces treated with N-methyl-3-aminopropyltrimethoxysilane.

### 4.5.2. Electrokinetic Effects of pH

Surfaces of metals, metal oxides, and silicates in equilibrium with atmospheric moisture are made up of hydrated hydroxyl groups. An electrokinetic surface potential (zeta potential) is imposed on these surfaces in an aqueous environment as a function of the $H^+$ and $OH^-$ ion concentrations. The pH at which the zeta potential of a surface is zero is the isoelectric point of the surface (IEPS). In an aqueous medium of a higher pH, the surface will be anionic, while at lower pH it will be cationic. Parks[23] has attempted to correlate the IEPS values for a large number of oxides by a simple electrostatic model involving cationic size and charge. Most surfaces cannot be described by a single IEPS, but depend on the oxidation state of the cation, the degree of hydration of the oxide surface, and the presence of impurities. Oxides with a higher degree of oxidation (higher cation valence) have a lower IEPS; for example, ferrous versus ferric oxides of Table 4.3. Completely hydrated surfaces yield a higher IEPS than freshly calcined surfaces.

Silane triols of neutral organofunctional silanes, $RSi(OH)_3$, have maximum stability at pH 2–4, which suggests that the isoelectric point of organofunctional silanols would be about 3. Simple electrostatic attraction of coupling agent silanols for hydrated oxide surfaces cannot be a major factor in bonding across the interface, since such silanols are more effective on acidic oxide surfaces than on basic oxide surfaces,[24] even though electrostatic forces would give repulsion of like charges on coupling agent and surface. In addition to electrostatic attraction or repulsion, coupling agent silanols may also react with hydroxylated surfaces through hydrogen bonding and through covalent oxane bonds.

Amino-organofunctional silanes exist in water as internal zwitterions (see Section 3.3.1). Electrokinetic effects may therefore be very important

**Table 4.3.** Isoelectric Points of Oxide Surface (IEPS) in Water

| Oxide surface | IEPS | Oxide surface | IEPS |
|---|---|---|---|
| $Ag^+$ | >12.0 | $Zn^{2+}$ | 9.0 |
| $Mg^{2+}$ | 12.2 | $Al^{3+}$ | 9.1 |
| $Fe^{2+}$ | 12.0 | $Fe^{3+}$ | 8.5 |
| $Co^{2+}$ | 11.3 | $Cr^{3+}$ | 7.0 |
| $Ni^{2+}$ | 11.0 | $Zr^{2+}$ | 6.5 |
| $Pb^{2+}$ | 10.3 | $Ti^{4+}$ | 6.0 |
| $Cd^{2+}$ | 10.3 | $Sn^{4+}$ | 4.5 |
| $Be^{2+}$ | 10.1 | $Mn^{3+}$ | 4.2 |
| $Cu^{2+}$ | 9.1 | $Si^{4+}$ | 2.2 |

in determining the orientation of such silanes deposited on oxide surfaces from water. It should make considerable difference in performance of silane coupling agents whether the organofunctional group or the silanol group was oriented toward the oxide surface.

Performance of amino-organofunctional silanes on fiberglass in polyester and epoxy composites was improved when the silane was applied to glass from acidified aqueous solutions, rather than from normally alkaline solutions.[25]

A nonionic methacrylate ester-functional silane (MPS) and a cationic vinylbenzyl functional silane (CSS) were compared as coupling agents on silica in polyester composites. Silanes were applied to silica from aqueous solutions in a pH range of 2 to 12.[26] Mechanical properties of composites were influenced by orientation of silanes on the silica surface as predicted by electrokinetic effects. Silica treated with the cationic silane is most sensitive to the pH of application, showing a steep improvement as the pH is reduced from 12 to 2. The nonionic silane is much less sensitive to pH of application, but also is most effective at a pH of 2 (Figure 4.5).

Because silica has an IEPS of about 2 to 3, its surface takes on increasingly strong negative charges as the pH is raised from 2 to 12. The cationic silane, then, will have increasingly strong electrokinetic tendency to orient "upside down," with the amine toward the silica surface. As the

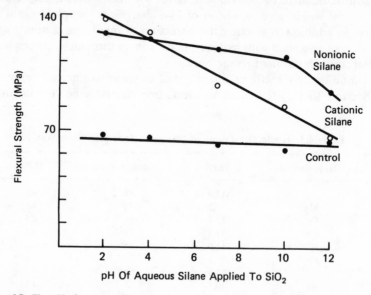

**Figure 4.5.** The pH of aqueous silane treatment (1%) on $SiO_2$ vs. flexural strength of polyester castings. Minusil (5 $\mu$m) in 30 g of resin.

pH of the treating bath is adjusted toward 2, the silica becomes less negative while the amine becomes more positive. Below the isoelectric point the silica also becomes positive and repels the amine, causing "right-side-up" orientation. The optimum pH should be at the isoelectric point, because at this point there is minimum charge on both silica silanol and coupling agent silanol, allowing them to approach each other for hydrogen bonding and siloxane formation. Similar tests on E-glass fibers showed optimum perform- ance of cationic silanes at pH 5, with poorer performance at pH 2 or 8. The isoelectric point of E-glass surface has not been reported, but a value of 5 seems reasonable from its surface composition (Table 4.2).

Measurements of zeta potential of silylated alumina surfaces were used by Kulkarni and Goddard[27] to study the nature of silane adsorption on the surface. Chemisorption of a neutral organofunctional silane[27] (MPS, Table 2.2) will result in the loss of ionizable groups of alumina, and hence will lead to a decrease in the surface charge. A surface completely silanated with completely condensed silane coupling agent will have no ionizable sites left and will show no surface charge, irrespective of the solution pH. Silanation with an incompletely condensed coupling agent will result in substitution of silanol sites for aluminum hydroxide sites and will shift the isoelectric point to a lower pH. If the adsorbed silane is hydrolytically unstable, desorption of silane in water will allow the isoelectric point to drift back to the value for untreated alumina.

The electrokinetic properties of alumina were determined by measuring streaming potentials at various pH's through a plug of powdered mineral. The zeta potential profiles of treated alumina compared with untreated alumina are depicted in Figure 4.6. Alumina was treated with the silane by either dry blending or by slurrying from solvent. Several observations should be noted.

1. The point of zero charge for untreated alumina was observed at 8.7, which is comparable to literature values of IEPS for alumina.[28]
2. Dry treatment of alumina with silane caused only slight change in the zeta potential profile, indicating that chemisorption did not occur on most of the surface.
3. Slurry-treated alumina showed much more silanol character than dry-treated or untreated alumina, indicating that much chemisorp- tion must have occurred. The zeta potential is less pH sensitive below pH 7, indicating some degree of silanol condensation.
4. Slurry-treated alumina showed no drift in zeta potential during 200 minutes in water, indicating that there was no appreciable desorption of the silane from the surface.

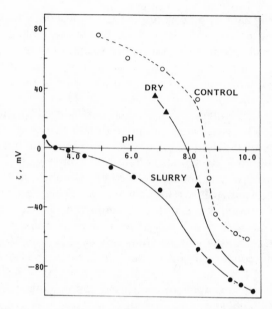

**Figure 4.6.** Zeta potential of alumina (T-61) treated with silane D (Table 6.1).

An interesting extension of this work would be to treat minerals with cationic- or anionic-organofunctional silanes to cause drastic changes in electrokinetic properties.

### 4.5.3. Contact Angle Measurements

Contact angle measurements of standard liquids on silane-treated surfaces were used by Bascom[28] and by Lee[29] to calculate critical surface tensions characteristic of the various silylating agents (Section 1.5.2, Table 1.11).

Silylated surfaces, dried under mild conditions, showed a large hysteresis in contact angle, the advancing angles being much larger than receding angles. This observation indicates that solvents penetrated the coupling agent layers, which must have an open structure in the vicinity of the glass surface.

Lee observed that the apparent critical surface tension of a vinylsilylated surface increased from 25 to 40 mN/m when dried at 80°C for 15 hr. This change was at first attributed to oxidation of the vinyl group, but more likely was the result of silanol condensation that altered the orientation of vinyl groups and exposed some of the siloxane backbone.

Generally, what has been evaluated is the critical surface tension, $\gamma_{crit}$, which rests on an operational definition and is without thermodynamic basis: it is the highest surface tension of a liquid that will spread spontaneously on a given surface. It is evaluated in the following way. A number of nonwetting liquids, over a range of surface tensions, are placed on a given surface, and the contract angle $\theta$ is evaluated. A plot is then constructed, with $\cos \theta$ as the ordinate and $\gamma_L$ as the abcissa. A straight line is expected, intercepting the ordinate ($\cos \theta = 1$) at $\gamma_{crit}$. In practice, however, a linear Zisman plot is rarely obtained on a silylated glass surface.

The problem appears to stem from not using a homologous series of liquids to make the Zisman plot. Thus, neither the dispersive component of surface tension, $\gamma_L^d$, nor its polar component, $\gamma_L^p$, is constant for the liquids used, nor do they vary linearly. Kaelble et al.[30] introduced a graphical method of obtaining dispersive $\gamma_s^d$ and polar $\gamma_s^p$ components of surface tension from contact angle measurements with liquids of known dispersive and polar components of surface tension.

Lee[31] and Sacher[32] used Kaelble's graphical method to obtain $\gamma_s^d$ and $\gamma_s^p$ for a number of silylated glass surfaces. Neglecting any small inductive component of surface tension, the sum of polar and dispersive surface tensions gives a total surface tension $\gamma_s$ that is compared with the operational $\gamma_{crit}$ previously reported. Calculated solubility parameters for compounds where R—H corresponds to the R—Si≡ silylating agent showed a rough linear correspondence with observed critical surface tensions of treated surfaces.[33] Solubility parameters may also be divided into components representing polar $\delta_p$, dispersion $\delta_d$, and hydrogen bonding $\delta_H$ forces between molecules.

Solubility parameter data of Hansen[34] for R—H compounds are compared with surface energy data of Lee for glass treated with corresponding silanes R—Si(OCH$_3$)$_3$ in Table 4.4. The rather high polar component of surface tension corresponding to nonpolar silanes suggests that orientation

**Table 4.4.** Surface Properties of Siloxane Films on Glass

| Organic group on silicon | Surface tensions (mN/m) | | | | Solubility parameters[a] | | |
|---|---|---|---|---|---|---|---|
| | $\gamma_{crit}$ | $\gamma_s^d$ | $\gamma_S^p$ | $\gamma_S$ | $\delta_d$ | $\delta_p$ | $\delta_H$ |
| Methyl | 22.5 | 16.0 | 8.0 | 24.0 | 6.3 | 0 | 0 |
| Propyl | 28.5 | 18.1 | 13.6 | 31.7 | 6.8 | 0 | 0 |
| 1,1,1-Trifluoropropyl | 33.5 | 16.0 | 18.7 | 34.7 | — | — | — |
| 2-Cyanoethyl | 34.0 | 24.0 | 11.5 | 35.5 | 7.5 | 7.0 | 2.7 |
| 3-Acetoxypropyl | 37.5 | 16.8 | 16.8 | 33.6 | 7.7 | 2.2 | 3.3 |
| Phenyl | 40.0 | 14.4 | 18.8 | 33.2 | 9.0 | 0 | 1.0 |

[a] Data of Hansen[34] (in Hildebrands) for R-H compounds corresponding to silanes R—Si(OCH$_3$)$_3$.

of the siloxane film on glass allowed siloxane bonds and possibly silanol groups to contribute significantly to surface properties.

Bascom[35] reported that the apparent $\gamma_{crit}$ of silica treated with ethyltriethoxysilane was reduced from 33 to 26 mN/m by adding an amine catalyst to the silane treatment to promote condensation of silanol groups.

Surface tension data of glass that had been treated with silanes of Table 2.2 were reported by Lee[31] and by Sacher.[32] The generally poor agreement of their data (Table 4.5) suggests that a rather broad range of surface tensions may be obtained with any silane, depending on the method of application, degree of silanol condensation, etc. Additional change is observed if the organofunctional group on silicon undergoes chemical reaction in the process. For example, hydrolysis of the epoxyfunctional silane in acid solution gives a glyceryl ether-functional silicone film that provides a very high polar component to surface tension.

In a review of the surface activity of silicones, Owen[36] observed that dimethylsiloxanes on a glass surface must be cross-linked by curing at >200°C to get reproducible contact angle measurements. Partial exposure of siloxane backbone gives a $\gamma_{crit}$ of 24 mN/m, which is higher than the liquid surface tension of polydimenthylsiloxane (20.4 mN/m). Consequently, the polymer will spread over its own adsorbed film. This creep of silicone fluids can be a practical problem if unreacted fluid migrates to electrical contacts. Conversely, it can be an advantage in achieving complete

**Table 4.5.** Surface Free Energies of Functional Polysiloxanes on Glass (N/m at 20°C)

| Organofunctional group on silicon | (Table 2.2) | Ref. | $\gamma_s^d$ | $\gamma_s^p$ | $\gamma_s$ | $\gamma_{crit}$ |
|---|---|---|---|---|---|---|
| Vinyl | A | 31 | 19.0 | 8.2 | 27.2 | 25.0 |
| Chloropropyl | B | 31 | 18.0 | 15.0 | 33.0 | 40.5 |
|  |  | 32 | 26.5 | 11.2 | 37.7 | 41.0 |
| Glycidylether | C | 31 | 8.7 | 40.1 | 48.8 | 40.0 |
|  |  | 32 | 26.8 | 19.6 | 46.4 | 39.0 |
| Glycidylether (acid hydrolyzed) | C | 32 | 23.0 | 32.3 | 55.3 | 44.6 |
| Methacrylate | D | 31 | 14.1 | 22.6 | 36.7 | 28.0 |
|  |  | 32 | 26.1 | 12.8 | 38.9 | — |
| Primary amine | E | 31 | 17.2 | 17.0 | 34.2 | 35.0 |
|  |  | 32 | 26.1 | 17.2 | 42.8 | 37.5 |
| Diamine | F | 31 | 23.0 | 7.8 | 30.8 | 33.5 |
|  |  | 32 | 26.9 | 15.5 | 42.4 | 36.0 |
| Mercaptopropyl | G | 31 | 20.4 | 22.6 | 41.0 | 43.0 |
|  |  | 32 | 25.5 | 14.7 | 40.2 | 33.6 |

surface coverage in metal protection, pigment surface treatment, and mold release. A commercial blend of high-viscosity silicone fluid with a lubricating particulate solid, such as mica, provides the combination of slip, air-bleed release, and thermal stability needed as mold release agent in modern tire presses. It is therefore somewhat paradoxical to think of organofunctional silanes as adhesion promoters. There is no conflict in fundamental surface chemistry, since abhesion and adhesion are closely related. Both require a stable, adhered surface layer. In release agents, the surface layer of silicone is inert to the contacting organic polymer, while the surface layer of silane coupling agent is effective in bonding to the organic matrix.

## 4.6. Chemical Reactions at the Interface

### 4.6.1. General

It has often been proposed that interfacial bonding due only to dispersion forces is theoretically sufficient to provide adhesion of organic polymers to mineral substrates that exceed the cohesive strengths of the matrix resins. Under ideal conditions, then, it would be sufficient to provide complete wetting between resin and reinforcement to obtain optimum strengths of composites. Conditions, in practical applications, are never ideal, and the matrix resin must compete with casual contaminants, low-molecular-weight fractions, and water for the interface. Water is the most damaging competing agent for hydrophilic mineral surfaces. Strong hydrogen bonding between water and hydroxyls on the mineral surface compete favorably with any possible bond between organic resin and the surface. At ambient conditions, such mineral surfaces are covered with at least a monolayer of water.

Alkoxysilanes are capable of reacting with surface moisture to generate silanol groups which also may form strong hydrogen bonds with the hydroxylated surface. In addition, they are capable of reacting with surface hydroxyl groups to form covalent oxane bonds with the mineral surface. Much indirect evidence suggests that all of these possible reactions do, in fact, occur between silane coupling agents and common mineral fillers and reinforcements.

It was relatively easy to demonstrate that appropriate organofunctional silanes reacted with thermosetting resins to link through covalent bonds. It was a little more difficult to explain bonding of thermoplastic polymers to silanes where no chemical reaction could be expected. Similarly, the bonding of silane coupling agents to silica and silicate minerals seemed logical by the unwritten chemical principle of "like-on-like," but the nature of bonding between silanes and metal oxide surfaces is still not completely understood (see Section 4.6.3).

### 4.6.2. Bonding to Siliceous Surfaces

Stark et al.[37] used radioactive tracers to demonstrate that trimethylsilanol gave only physical absorption on glass or silica unless a silanol condensation catalyst was present. Trimethylchlorosilane combined chemically with an E-glass surface, but was readily displaced by boiling water. Stability of coupling agent layers on glass and silica in the presence of water was measured by radio-tracer techniques (Section 4.4.4), but specific chemical reactions could not be identified by this technique.

Infrared spectroscopy is a well-established method of studying the nature of absorption on high-surface-area powders like fumed silica. White[38] observed that methylchlorosilanes interact physically and reversibly with silanol groups of a dry silica surface at room temperature. Irreversible chemisorption took place when silica was treated with chlorosilanes in $CCl_4$ solution and dried at 150°C. Chemically bonded methylsilanes on silica were stable to heat and water vapor at 500°C, but were destroyed in air at 400°C.

Duffy[39] found that essentially no reaction occurred at 25°C between ethylsilicate and silica, while at 100°C, traces of byproducts were identified by gas chromatography. At 165°C, a condensation reaction was observed with both hydrated and predried silica. Infrared spectroscopy indicated that silanol sites on silica took part in a reaction that formed ethanol and new siloxane bonds.

The gas-phase reaction of methyltrimethoxysilane with silica gel in the range from room temperature to 200°C was studied by Hertl[40] using IR spectroscopy. At room temperature the silane was physically adsorbed and could be removed by pumping. At 120–200°C the adsorbed silane reacted chemically with isolated silanol groups on the silica with third-order reaction kinetics and an activation energy of 30.5 kcal/mol. For every three isolated silanol groups on silica that reacted, 1.56 formed SiOSi bonds with the methoxysilane, 0.45 formed hydrogen bonds with unreacted methoxy groups of silane that had formed siloxane bonds with adjacent silanol groups, and 1.08 formed $SiOCH_3$ bonds with liberated methanol. These bonds were stable to water vapor at room temperature and resisted thermal changes to 300°C in vacuum. Reactivity of silica was increased by predrying at higher temperatures, indicating that reaction was directly between isolated silanol groups and the methoxysilane, and did not require prehydrolysis of the methoxysilane with surface water.

Alksne et al.[41] attempted to study the nature of siloxane films deposited on E-glass fibers from aqueous solutions of silane coupling agents. They were able to observe some hydrolysis of CN groups of beta-cyanoethyl-triethoxysilane by absorption typical of the carbonyl group, but found the sensitivity of dispersive infrared spectra to be insufficient to show details of chemical structures at the interface.

Application of IR spectroscopy to the study of coupling agent–glass reactions is limited by the strong background absorption bands of bulk glass. Koenig and Shih[42] used argon-ion-laser excited Raman spectroscopy to study reactions of vinyltriethoxysilane (VTS) on glass fibers and micro-beads. The weak Raman scattering of bulk glass allowed them to observe formation of siloxane bonds at the glass surface, as indicated by absorption at 788 cm$^{-1}$. After treated glass was boiled for 2 hr in water, the frequency shifted to 783 cm$^{-1}$, which is characteristic of siloxane homopolymer. When redried at 110°C, the treated glass again showed the line at 788 cm$^{-1}$. These data were interpreted as evidence that the coupling agent condensed with glass to form covalent siloxane bonds across the interface, but that the reaction was reversible in the presence of water. The weak intensity of bonds for siloxane and silanol identification is a limiting factor in using laser Raman spectroscopy for studying the interface in greater detail.

A strong UV resonance Raman line at 1545 cm$^{-1}$ was used by Ishida et al.[43] in studying a phthalocyanine-functional silane coupling agent on glass fibers and on high-surface-area silica. The remarkable enhancement of silane signals allowed this technique to identify coupling agent layers of less than a monolayer thickness on glass or silica.

It has long been a goal to obtain infrared spectra of coupling agents on glass-fiber surfaces. Unfortunately, the sensitivity and selectivity of dispersion instruments are insufficient to obtain the spectral information about the coupling agent whose concentration is small compared with the amount of bulk glass. FT-IR is an extremely powerful technique for studying aqueous solutions of the silane and of a coupling agent–high-surface-area silica powder system. Digital subtraction of spectra using a dedicated minimcomputer gave direct quantitative measurements, particularly on the SiOH groups, as well as on the other organic groups of silane coupling agents. As a consequence, the structural changes of the coupling agents on the glass surface could be followed.

Ishida and Koenig[44] used a FT-IR spectrophotometer (Digilab FTS-14) with a magnetic tape unit for their study of vinylsilanols in solution. A solution sample was placed between two AgBr windows with a 3-$\mu$m-thick Mylar spacer. Spectra were coadded 200 and 400 times at a resolution of 2 cm$^{-1}$. Wave number accuracy was better than 0.01 cm$^{-1}$. The spectrum of a silane solution was stored in the memory unit. Another spectrum of a solvent (distilled water or an alcohol–water mixture) was taken and stored in the same fashion. Then the spectra were recalled and the contribution of the solvent absorption was digitally subtracted from the spectrum of the silane aqueous solution, using the water band and the C—O stretching mode of the alcohol as internal thickness bands (Table 3.5).

Crystalline phenysilanetriol was studied in the same manner[45] in order to obtain firm assignment of frequencies for studying silanol reactions at mineral surfaces. Using these data, Ishida and Koenig studied the structure and chemistry of silane coupling agents on high-surface-area silica[46] and on E-glass fibers[47,48] (Figure 4.7). The reversible nature of silanol condensation and rehydrolysis on E-glass fibers was demonstrated.[49] The molecular organization of silane coupling agents on glass was studied by comparing vinylsilanols with pure cyclohexylsilane triol on glass surfaces. The composition of aqueous aminopropylalkoxysilanes and their orientation on surfaces have been the subject of much speculation with little direct evidence. A FT-IR study of this subject has removed some of the mystery about the action of this important class of silane coupling agents.[50] The original papers should be consulted by those interested in details of spectral interpretation.

FT-IR data indicated that neutral organofunctional trialkoxysilanes form silane triols in water, which condense slowly at pH 4–6 to oligomeric siloxanols. When applied to glass from dilute aqueous solution, there is

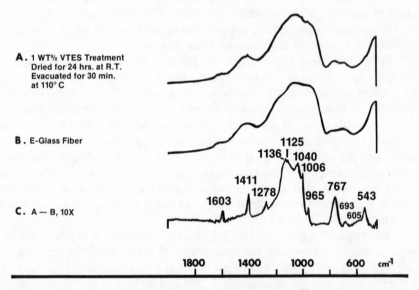

**Figure 4.7.** FT-IR spectra on E-glass fiber–coupling agent system.[46] A. E-glass fiber treated with 1% by weight vinyltriethoxysilane and dried for 24 hr at room temperature with subsequent evacuation for 30 min at 100°C. B. E-glass fiber heat cleaned at 500°C for 24 h. C. Polyvinyl-siloxane on the E-glass fiber with the contribution of the glass excluded. (Courtesy of Academic Press.)

considerable orientation of silanol groups toward the glass. The rate of condensation of vinylsilanols was faster on glass than in the bulk (Figure 4.8). This may be due to favored orientation and organization within the silane layer. The methacrylate-functional silane (MPS) on glass showed less hydrogen-bonded carbonyl than in precipitated polysiloxanol, suggesting a predominant head-to-head adsorption of coupling agent layers.

Aminopropyltrialkoxysilanes (APS) were studied in aqueous solution and as films on silica and E-glass fibers.[51] Alkoxy groups hydrolyze rapidly in water. Siloxane bonds appear immediately along with a shift in the $NH_2$ deformation band from $1600 \, \text{cm}^{-1}$ to $1575 \, \text{cm}^{-1}$. After complete drying, the absorption in APS films shifts back to $1600 \, \text{cm}^{-1}$. The band at $1575 \, \text{cm}^{-1}$ corresponds to hydrogen-bonded $NH_2$, while the band at $1600 \, \text{cm}^{-1}$ corresponds to free $NH_2$ groups. Thus, hydrolyzed APS exists in two structural forms, probably in equilibrium: One is an open, extended nonring structure and the other is a cyclic hydrogen-bonded amine structure:

$$
\begin{array}{ccc}
-O & & CH_2CH_2 \\
& \diagdown \diagup & \diagdown \\
& Si & \quad CH_2 \\
& \diagup \diagdown & \diagup \\
-O & OH\text{---}NH_2 &
\end{array}
$$

A weak absorption band at $2150 \, \text{cm}^{-1}$ suggests a partial proton transfer $SiO\text{---}H\text{---}NH_2^+$ in the ring structure. This type of ring structure explains the stability of aqueous solutions of APES under conditions that would normally give immediate precipitation of a mixture of propyltrimethoxy-silane and amine in water.

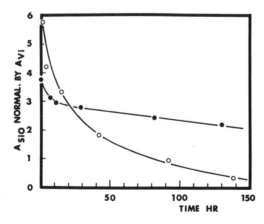

**Figure 4.8.** Infrared intensities of the band due to the residual SiOH groups at various drying times. Open-circle and closed-circle represent with and without glass fiber, respectively.

The hydrolyzed aminosilanes adsorb from dilute solution onto high-surface-area silica as monolayers with strong hydrogen-bonding interactions of amine groups with the silica surface. On the other hand, films on E-glass are thicker multilayers with the amino groups held in intramolecular ring structures. The relative proportion of the two major structures is probably related to the pH of the treating solution and may be controlled to determine the availability of amine groups for coupling to matrix resins in composites.

Hydrolytic stability of silane coupling agent films on E-glass could also be observed by FT-IR.[52] Multilayers of coupling agent deposited from water show a high degree of condensation of silanols to siloxanes during drying under vacuum at 110°C. Desorption in water at 80°C is different for different organofunctional silanes, but generally shows a threshold period during which there is no desorption (Figure 4.9). Silanol groups are formed during the threshold period, and when the size of the resulting siloxanol oligomer is small enough, desorption in water takes place. Films of cyclohexylsiloxane on glass show remarkable resistance to desorption in water. Silanol formation reached an equilibrium level in about 1000 hr at 80°C, but no desorption occurred in 5000 hr. Films of vinylsiloxanes on glass showed silanol formation in water during 500 hr at 80°C without desorption, and after this threshold period showed slow desorption during the next 1000 hr.

Rosen and Goddard[53] observed similar differences in desorption of alkylsiloxane films from treated fillers by measuring surface tension of water in contact with the fillers. Amyl-, butyl-, and methacryoxypropylsiloxane films showed moderate desorption from alumina trihydrate (ATH) within a few hours in water, while vinyl-, propyl-, phenyl-, hexyl-, and octylsiloxane films showed little desorption from ATH under the same conditions. These observed differences are probably due to differences in the structure of

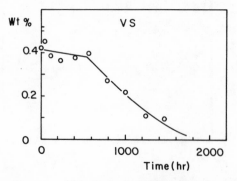

**Figure 4.9.** The desorption curve of VS adsorbed on E-glass fibers from 4% by weight aqueous solution, at 80°C in water.[49] (Courtesy of John Wiley & Sons.)

siloxane films. Siloxanes having small alkyl substituents (methyl, vinyl, propyl) condense to highly cross-linked insoluble siloxane films. Siloxanes having larger organic groups on silicon are more highly cyclized and less highly cross-linked, and therefore are more susceptible to desorption by water. Higher hydrocarbon substituted (hexyl, phenyl, cyclohexyl, octyl) siloxanes are so insoluble in water that even lightly cross-linked cyclic polysiloxanes are not desorbed in the first few hours by water. Rosen and Goddard did not expose treated fillers to water long enough to observe a threshold period of desorption as observed by Ishida and Koenig.

Siloxane films on mineral reinforcement are not dependent on homocondensation reactions to develop sufficient cross-linking to provide a water-resistant interphase region in composites, but are expected to also coreact with the matrix resin. Thus, Ishida and Koenig[52] observed a threshold time of 500 hr for desorption of siloxane films from E-glass, but after a MPS film was copolymerized with styrene monomer, it showed no desorption in 80°C water in 1500 hr (Figure 4.10).

Photoacoustic spectra correspond closely to optical absorption spectra of the sample, but may provide some advantage in studies of molecular structures on optically opaque surfaces. Leyden[54] et al. used photoacoustic techniques to study metal chelates immobilized on silica gel surface through organofunctional silanes. Photoacoustic techniques show promise in identifying keto/enol ratios of beta diketones on the surface, and may differentiate between mono- and bis-complexes of the bound chelating agents with metal ions. New instruments with laser excitation, having near-IR capabilities, may extend this work into the more useful range of molecular structures.

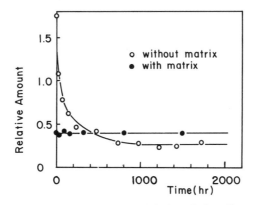

**Figure 4.10.** The desorption curve of γ-MPS deposited on E-glass fibers and copolymerized with styrene. The treated fibers were washed with $CS_2$ three times before use in order to dissolve the deposited polystyrene homopolymer.

### 4.6.3.  Bonding to Metal Oxides

Monomolecular layers of silane coupling agents on aluminum have been observed by inelastic electron tunneling spectroscopy[55] (IETS). Very clear spectra of aminofunctional silanes on aluminum showed the presence of unhydrolyzed alkoxy groups on silicon (silanes were applied from anhydrous benzene) and shifts in absorption of the amino group attributed to reaction with the metal surface. IETS may have an advantage over ESCA (and other high-vacuum techniques) in studying the effects of moisture or other disrupting chemicals on orientation and bonding at the metal surface.

Furukawa et al.[56] studied the vapor-phase adsorption of APS on aluminum by IETS. Initial adsorption seemed to be through the amino group. Upon exposure to moist air much of the silane was desorbed while the residue appeared to undergo hydrolysis (of ethoxy groups from silicon), and a rearrangement with the silanol groups now bonded to the aluminum oxide surface.

IR spectroscopy has been used successfully to study adsorption on high-surface-area powders, and it has long been known that this type of spectroscopy could, in principle, be used to investigate thin organic films formed on bulk metal substrates if the necessarily weak infrared absorption of such films could be suitably enhanced and detected. In the past, efforts to enhance the adsorption have been made by increasing the geometrical path length length of the infrared rays in the film by making multiple reflections of the incident radiation at various angles (e.g., 30°–60°) from opposing samples. Bascom[57] obtained attenuated total-reflectance (ATR) infrared spectra of aminopropylsilane (APES) adsorbed into germanium from cyclohexane and from water. Spectra of films formed in these solvents were identical, indicating that the structures of the films were similar. Spectra obtained from films adsorbed from cyclohexane were, however, somewhat more intense, indicating that these films were thicker than those adsorbed from water. Bascom suggested that a broad band near 3000 cm$^{-1}$ in the spectra of the adsorbed films was the result of hydrogen bonding of the amine groups that normally absorb near 3350 cm$^{-1}$. When light is incident on a highly reflecting metallic surface at small angles, the incident and reflected waves combine to form a standing wave having a node very near the surface of the metal. Accordingly, the electric field has zero amplitude at the surface of the metal, and it cannot interact with any molecules adsorbed on the surface. As a result, this reflection arrangement is not expected to be effective for small incidence angles.

Greenler[58] reconsidered the problem of obtaining the infrared spectra of organic films formed on metallic surfaces. His results showed that only for radiation polarized parallel to the plane of incidence and only when making an angle of incidence that is only a few degrees less than 90° will the incident and reflected rays combine to establish a standing wave with appreciable amplitude at the surface of the metal to be capable of intereacting with adsorbed molecules. This prediction of high-absorption intensity for parallel polarized radiation and high angles of incidence is not based on the geometrical path length of the IR spectroscopy, and Greenler and associates have suggested it be called reflection-absorption (R-A) spectroscopy. Boerio et al.[59] used R-A spectral techniques in studying the adsorption of APS on polished iron and aluminum, and of methacryloxypropyltrimethoxysilane (MPS) on polished copper surfaces.

Films of APS deposited on iron mirrors at the natural pH (10.4) were composed of highly hydrolyzed oligomers, and were characterized by infrared bands near 1550 and 1480 cm$^{-1}$ that were assigned to $NH_2$ deformation modes in amino groups that were strongly hydrogen bonded to anionic groups. Further studies of APS films deposited over a wide pH range showed a transition at pH 9.5. Films formed at low pH were composed of polysiloxanes in which the amino groups form cyclic, internal zwitterions with silanol groups as shown by weak bonds near 1600 cm$^{-1}$ and 1500 cm$^{-1}$. Films deposited at high pH showed intense bonds at 1600 cm$^{-1}$ and 1500 cm$^{-1}$, believed to be related to interaction of amine groups with the iron oxide surface.

It may be noted that the isoelectric point of the oxidized surface of iron is near a pH value of 8.5. The oxide will, therefore, be positively charged in an aqueous environment at a pH less than 8.5, due to the adsorption of protons, and negatively charged at a pH greater than 8.5, due to ionization of surface hydroxyl groups. When APS is absorbed on the oxidized surface of iron at a pH of 9.5 or greater, there may be a specific interaction between the ionized iron hydroxyls and the amino groups.

Performance of APS films on iron as adhesion promoter should show differences when deposited from acid or alkaline pH, because at pH greater than 9.5 the silane would have an "upside-down" orientation on the surface. To investigate this possibility, single lap shear adhesive joints were prepared and tested. As expected, the samples pretreated with aqueous solutions of APS all broke at much higher loads than the untreated control samples. Moreover, samples pretreated at the highest pH (12.0) failed at substantially lower loads than samples treated at a pH of 9.5 or less.

If the transition observed near pH = 9.5 for APS films adsorbed onto iron from aqueous solutions is related to the isoelectric point of the substrate, the transition will be expected to occur at a different pH if the substrate is changed. Accordingly, the absorption of APS onto aluminum was investigated. The results obtained for aluminum substrates were very similar to those obtained for iron substrates, with a transition again observed near pH = 9.5. This may indicate that the transition is only a property of the silane and is unrelated to the substrate. However, the isoelectric point of an oxidized aluminum surface is at about pH = 9.4, and this may simply be too close to the isoelectric point of an oxidized iron surface (~8.5) to enable any difference to be noted. The similarity of coupling agent spectra on iron and aluminum mirrors suggests that any unique iteraction with the metal oxides is not strong.

Adsorption of MPS onto copper mirrors from 1% solutions in cyclohexane showed an intense absorption near 1720 cm$^{-1}$ that may be assigned to the stretching mode of the C=O bond.[60] The weaker band near 1635 cm$^{-1}$ is undoubtedly due to the C=C stretching mode. The intensity of the band near 1720 cm$^{-1}$ indicates that the films adsorbed on these coupons were several molecular layers in thickness. Rinsing the coupons for 1 min in cyclohexane resulted in a marked decrease in the intensity of the band near 1720 cm$^{-1}$, indicating that much of the adsorbed MPS was only weakly bound. Further increases in the rise time did not result in a further decrease in intensity of the band near 1720 cm$^{-1}$.

It may be concluded that films formed by adsorption of MPS onto copper from cyclohexane are several molecules thick. Most of this material is loosely bound, however, and may easily be desorbed to leave a thin film that is perhaps one or two molecular layers thick. Careful examination of the band near 1720 cm$^{-1}$ indicated the presence of a shoulder on the low-frequency side of the band. Similar shifts to lower frequencies have been observed for the C=O bond. This result is especially important because Lee[29] has concluded from wettability studies that MPS is adsorbed on glass with the C=O bond oriented toward the surface.

Sung and Sung[61] found two major advantages in choosing single-crystal sapphire as a model for an aluminum surface. The first advantage is the transmittance of sapphire in the IR radiation, up to 6.7 (~1500 cm$^{-1}$), which makes it possible to obtain transmission IR spectra showing the chemical and structural changes of silanes and possible interactions between silane and aluminum oxide. The second advantage is that when the silane film is very thin, sapphire plates can be used as an internal reflection plate, because the refractive index of sapphire is much greater than that of silane film.

Transmission and internal reflection spectra of APS films applied from water clearly showed that the films on sapphire were polysiloxane networks. A shift in absorption for NH$_2$ was again observed and attributed to the presence of protonated nitrogen, but films were too thick to allow interpretation of interface structures.

A film deposited from dilute acidified solution of vinyltriethoxysilane also showed a siloxane network in which cross-linking was not complete, since silanol (SiOH) groups were still present.

Boerio et al.[62] examined the effects of silane coupling agents on the stability of the oxide layer on aluminum. The air-formed oxide was quickly converted in 60°C water to weak crystalline pseudo-böhmite. Thin layers of silane on the oxide inhibited this rate of hydration, possibly through the formation of aluminosiloxane bonds.

The substance 3-aminopropyldimethylethoxysilane has only one alkoxy group, and it may therefore react with a substrate without the complication of transfacial Si—O—Si groups. Diffuse-reflectance FT-IR spectra of this substance on alumina and titania have recently been reported by Naviroj et al.[63] Spectra of treated and untreated titania were compared with the difference spectrum. A weak band at 950 cm$^{-1}$ was tentatively assigned to interfacial Si—O—Ti groups, and this was confirmed by using silica as the substrate and a titanate [isopropyltri(isostearoyl)titanate] as the coupling agent. The neat titanate has no bands around 950 cm$^{-1}$, but the difference spectra (treated minus untreated silica) fortunately reveals a peak at 950 cm$^{-1}$. The silane on alumina reveals a weak peak at 963 cm$^{-1}$ attributed to Si—O—Al groups.

Gettings and Kinloch[64] detected an ion of mass 100, assigned as FeSiO$^{+}$, in the SIMS spectrum of a steel surface treated with a 1% aqueous solution of GPS. This ion was absent when APS or a styrene-functional amine hydrochloride silane was employed. GPS gave joints of superior durability when immersed in water at 60°C for up to 1500 hr. Gettings and Kinloch claim that "this is strong evidence for the formation of a chemical bond, probably —Fe—O—Si— between the metal oxide and polysiloxane primer."

### 4.6.4. Bonding to Polymers

In addition to indicating the nature of mineral surface reactions of silane coupling agents, FT-IR also has given some indication of organofunctional group reaction with matrix resins. Untreated E-glass inhibits the polymerization of unsaturated polyester resins[4] as was also inferred from studies of cure exotherms of filled resins. When adsorbed on silica, both vinylsiloxane and methacryloxypropylsiloxane showed loss of C=C double bonds during cure with styrene-modified unsaturated polesters.[48] The methacrylate silane also indicated copolymerization on E-glass, but the vinyl silane remained unreacted. As shown in Figure 4.8, an E-glass surface catalyzes the condensation of silanols to such an extent that the vinyl siloxane is too highly cross-linked to interdiffuse with the polyester resin. The methacryloxypropylsiloxane remains soluble and fusible even when highly condensed due to its strong tendency to form cyclic siloxane structures.

Chiang and Koenig[65] studied the polymerization of an anhydride-cured epoxy resin in contact with fiberglass that had been treated with N-methyl-aminopropylsilane solutions. The silane caused an acceleration in appearance of ester linkages as well as amide linkages, with accelerated loss of unreacted anhydride. It was estimated that the silane-matrix interphase region had 5–10% greater cross-link density than the bulk of the epoxy matrix.

The chemical reaction of APS on glass with anhydride-cured epoxy resin was studied by Chiang and Koenig.[51] Although epoxy resin reacts directly with some of the primary amine of the coupling agent, a large proportion of the amine reacts with the anhydride (NMA) to form a cyclic imide. This proportion of aminofunctional silane and curing agent is lost in terms of bonding across the interface. A secondary aminofunctional silane reacted normally with epoxy and NMA during cure cross-linking. This suggests that the reactions between resin and silane provided a cross-linked interpenetrating network linking glass to the matrix.

The importance of interdiffusion and cross-linking in the interphase region was demonstrated by Chaudhury et al.[66] in bonding polyvinylchloride (PVC) plastisols to silane-primed germanium. Silicon concentration in the interfacial region was examined by XPS (X-ray photoelectron spectroscopy) depth profile analysis starting from the germanium layer (Figure 4.11).

Diffusion of siloxane into the plasticized PVC fiber was much greater if the primer was dried at 25°C rather than 175°C. Resistance of the bond to water, however, was optimum when the primer was partially cross-linked by drying at intermediate temperatures.

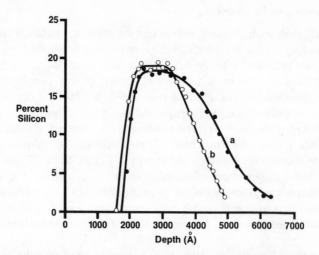

**Figure 4.11.** Atomic percent of silicon as a function of sputter depth of germanium bonded to a PVC plastisol through a silane primer, curve $a$ = primer dried at 25°C and curve $b$ = primer dried at 175°C. (Courtesy of BV Science Press, Utrecht, The Netherlands.)

Sung et al.[67] treated a sapphire surface with APS and then bonded molten polyethylene to it. Because polyethylene has few functional groups, it is likely that interaction between polymer and APS must be physical. Energy-dispersive X-ray analysis indicated that the silane diffused into the polyethylene, and peel strength measurement showed an increase with thickness of the APS layer up to that obtained from a 2% aqueous solution.

## References

1. D. E. Leyden, G. H. Luttrell, W. K. Nonedez, and D. B. Werho. *Anal. Chem.* **48**, 67 (1976).
2. D. E. Leyden and G. H. Luttrell. *Anal. Chem.* **47**, 1612 (1975).
3. G. D. Nichols, D. M. Hercules, R. C. Peek, and D. J. Vaughan. *Appl. Spectros.* **281**, 219 (1976).
4. H. O. Finkler and R. W. Murray. *J. Phys. Chem.* **83**(3), 353 (1979).
5. K. W. Willman, E. Greer, and R. W. Murray. *Nouv. J. Chim.* **3**(7), 455 (1979).
6. J. F. Cain and E. Sacher. *J. Colloid Interface Sci.* **67**(3), 538 (1978).
7. R. Wong. *J. Adhesion* **4**(2), 171 (1972).
8. G. D. Nichols and D. M. Hercules. *Appl. Spectrosc.* **28**(3), 219 (1974).
9. A. T. DiBenedetto and D. A. Scola. *J. Colloid Interface Sci.* **74**(1), 150 (1980).
10. B. D. Favis, L. P. Blanchard, J. Leonard and R. E. Prud'homme. *Polym. Compos.* **5**, 11 (1984).
11. D. J. Belton and A. Joshi. In *Molecular Characterization of Composite Interfaces*, H. Ishida and G. Kumer, Eds. p. 187, Plenum, New York (1985).
12. E. P. Plueddemann. *Applied Polym. Symp.* **19**, 75–90, Wiley, New York (1972).
13. N. M. Trivisonno, L. H. Lee, and S. M. Skinner. *Ind. Eng. Chem.* **50**, 912 (1958).
14. S. Sterman and H. B. Bradley. SPI, 16th Ann. Tech. Conf. Reinf. Plast. 8-D (1961).
15. G. E. Vogel, O. K. Johannson, F. O. Stark, and R. M. Fleischmann. SPI, 22nd Ann. Tech. Conf. Reinf. Plast. 13-B (1967).
16. D. J. Vaughan and R. C. Peek, Jr. SPI, 30th Ann. Tech. Conf. Reinf. Plast. 22-B (1975).
17. D. J. Tutas, R. Stromberg, and E. Passaglia. *SPE Trans.* **4**, 256 (1964).
18. W. D. Bascom. *Adv. Chem. Ser.* No. 87, 38 (1968).
19. M. E. Schrader, I. Lerner, and F. J. Doria. *Mod. Plast.* **45**, 195 (1967).
20. O. K. Johannson, F. O. Stark, G. E. Vogel, and R. M. Fleischmann. *J. Compos. Mater.* **1**, 278 (1967).
21. M. E. Schrader and A. Block. *J. Polym. Sci., Part C* **34**, 281 (1971).
22. F. J. Kahn. *Appl. Phys. Lett.* **22**(8), 386 (1973).
23. G. A. Parks, *Chem. Rev.* **65**, 127–198 (1964).
24. E. P. Plueddemann and G. L. Stark. SPI, 28th Ann. Tech. Conf. Reinf. Plast. 21-E (1973).
25. E. P. Plueddemann and G. L. Stark. SPI, 27th Ann. Tech. Conf. Reinf. Plast. 21-B (1972).
26. E. P. Plueddemann and G. L. Stark. *Mod. Plast.* **92**(3), 74 (1974).
27. R. D. Kulkarni and E. D. Goddard. SPI, 35th Ann. Tech. Conf. Reinf. Plast. 20-E (1980).
28. W. O. Bascom. *Adv. Chem. Ser.* No. 87, 38 (1968).
29. L. H. Lee. *J. Colloid Interface Sci.* **27**(4), 751 (1968).
30. D. H. Kaelble, P. J. Dynes, and E. H. Cerlin. *J. Adhesion* **6**, 23 (1974).
31. L. H. Lee. *Adhesion Science and Technology*, Vol. 9B, p. 647, Plenum, New York (1975).
32. E. Sacher. *Symposium on Silylated surfaces*, D. E. Leyden and W. Collins, Eds., p. 347, Gordon & Breach, London (1980).
33. E. P. Plueddemann. SPI, 25th Ann. Tech. Conf. Reinf. Plast. 13-D (1970).

34. C. M. Hansen. *J. Paint Technol.* **39**(511), 505 (1967).
35. W. D. Bascom. *J. Colloid Interface Sci.* **27**, 79 (1968).
36. M. J. Owen. *Ind. Eng. Chem. Prod. Res. Devel.* **19**(1), 97 (1980).
37. F. O. Stark, O. K. Johannson, G. E. Vogel, R. G. Chaffee, and R. M. Lacefield. *J. Phys. Chem.* **72**, 1248 (1968).
38. T. E. White. SPI, 20th Ann. Tech. Conf. Reinf. Plast. 3-D (1965).
39. J. V. Duffy. *J. Appl. Chem.* **17**, 35 (1967).
40. W. Hertl. *J. Phys. Chem.* **72**, 124 (1968).
41. V. I. Alksne, V. Th. Kronberg, and Ya. A. Eidus. *Makhanika Polimerov* **4**, 182 (1968).
42. J. Koenig and P. T. K. Shih. *J. Colloid Interface Sci.* **64**(3), 565 (1978).
43. H. Ishida, J. L. Koenig, and M. Kenney. SPI, 34th Ann. Tech. Conf. Reinf. Plast. 17-B (1979).
44. H. Ishida and J. L. Koenig. *Appl. Spectros.* **32**(5), 462 (1978).
45. H. Ishida and J. L. Koenig. *Appl. Spectros.* **32**(5), 467 (1978).
46. H. Ishida and J. L. Koenig. *J. Colloid Interface Sci.* **64**(3), 555 (1978).
47. H. Ishida and J. L. Koenig. *J. Colloid Interface Sci.* **64**(3), 565 (1978).
48. H. Ishida and J. L. Koenig. *J. Polym. Sci., Polym. Phys.* Ed. **17**, 615 (1979).
49. H. Ishida and J. L. Koenig. *J. Polym. Sci., Polym. Phys. Ed.* **18**, 233 (1980).
50. C. H. Chiang, H. Ishida, and J. L. Koenig. *J. Colloid Interface Sci.* **74**(2), 396 (1980).
51. C. H. Chiang and J. L. Koenig. SPI, 35th Ann. Tech. Conf. Reinf. Plast. 23-D (1980).
52. H. Ishida and J. L. Koenig. *J. Polym. Sci., Polym. Phys. Ed.* **18**, 1931 (1980).
53. M. R. Rosen and E. D. Goddard. *Polym. Sci. Eng. Sci.* **20**(6), 413 (1980).
54. D. E. Leyden, D. S. Kendal, L. W. Bruggrof, F. J. Pern, and D. Williams. Presented at Sept. 1980 Analytical Division ACS Meeting in Las Vegas.
55. A. F. Diaz, U. Hetzler, and E. Kay. *J. Amer. Chem. Soc.* **99**, 6781 (1977).
56. T. Furukawa, N. K. Eib, K. L. Mitta, and H. R. Anderson, Jr. *J. Colloid Interface Sci.* **96**(2), 322 (1983).
57. W. D. Bascom. *Macromolecules* **5**, 792 (1972).
58. R. G. Greenler. *J. Chem. Phys.* **44**, 310 (1966).
59. F. J. Boerio, F. J. Armogan, and S. Y. Cheng. *J. Colloid Interface Sci.* **73**(2), 416 (1980).
60. F. J. Boerio, S. Y. Cheng, L. Armogen, J. W. Williams, and C. Gosselim. SPI, 35th Ann. Tech. Conf. Reinf. Plast. 23-C (1980).
61. N. H. Sung and C. S. Paik Sung. SPI, 35th Ann. Tech. Conf. Reinf. Plast. 23-B (1980).
62. F. J. Boerio, R. G. Dillingham, and R. C. Bozian. SPI, 39th Ann. Tech. Conf. Reinf. Plast. 4-A (1984).
63. S. Naviroj, J. L. Koenig, and H. Ishida. *J. Adhesion* **18**, 93 (1985).
64. M. Gettings and A. J. Kinloch. *J. Mater. Sci.* **12**, 2511 (1977).
65. C. H. Chiang and J. L. Koenig. *J. Polym. Sci., Phys. Ed.* **20**, 2135 (1982).
66. M. K. Chaudhury, T. M. Gentle, and E. P. Plueddemann. *J. Adhesion Sci. Technol.* **1**(11), 29 (1987).
67. N. H. Sung, A. Kaul, I. Chin, and C. S. P. Sung. *Polymer Eng. Sci.* **22**, 637 (1982).

# 5 | Nature of Adhesion Through Silane Coupling Agents

## 5.1. Problems in Organic–Inorganic Composites

### 5.1.1. Shrinkage Stresses

Most thermosetting polymers undergo some shrinkage during cure. This may range from 6 to 10% for polyesters to 3 to 4% for epoxies. Such shrinkage in the presence of fillers or reinforcements sets up stresses at the polymer–resin interface. Bush[1] embedded strain gauge transducers in a polyester resin cured at 120°C and measured internal static pressures of 14 to 21 MPa (2000–3000 psi). Bulk mold shrinkage may be reduced to less than 1% by loading the polymer with particulate fillers, but interfacial stresses are merely transferred to the filler surfaces.

Additional stresses are set up at the interface between reinforcement and matrix resin during temperature changes due to differences in thermal expansion. Coefficients of linear thermal expansion of a few common plastics and minerals are shown in Table 5.1. The coefficient of thermal expansion of glass is about $6 \times 10^{-6}/°C$, while that of a typical rigid thermosetting plastic is about $60 \times 10^{-6}/°C$. One result is that unidirectional glass-fiber-reinforced laminates show different thermal expansion in the direction of the fibers than across the fiber axis. Stresses on single glass fibers embedded in clear thermosetting resins were measured by McGarry and Fyiwara[2] by a photoelastic technique. They observed about 14 MPa of radial pressure on the fiber and about 7 MPa of axial compression for every 50° of cooling below maximum cure temperature. Warming the composite obviously would reverse the direction of the forces. Compression of matrix resin around the fiber during cure and cooling sets up frictional resistance

**Table 5.1.** Linear Thermal Expansion of
Common Materials

| Material | Coefficient of thermal expansion $c \times 10^{-6}/°C$ |
|---|---|
| Silica glass | 0.6 |
| Corning 774 glass | 3.2 |
| Alkali (soft) glass | 9.0 |
| Alumina (alundum) | 8.7 |
| Aluminum | 23.0 |
| Steel | 10–14 |
| Graphite | 7.8 |
| Brass | 19.0 |
| Polyimide | 38–54 |
| Epoxy | 45–65 |
| Polyester | 55–100 |
| Phenolic | 60–80 |
| Polystyrene | 60–80 |
| Polypropylene | 100–200 |
| Cellulose acetate butyrate | 110–170 |
| Polyurethane resins | 100–200 |
| Silicone resin | 160–180 |

to fiber pullout in fracture, and in this way may be beneficial for the strength of composites.[3]

Reinforced thermoplastics do not undergo cure shrinkage during fabrication, but shrinkage stresses may be even more severe because these resins have high coefficients of thermal expansion, and they are usually fabricated at relatively high temperatures. Temperature cycling in the presence of moisture, moreover, would set up a strong pumping action to bring water to the interface.

### 5.1.2. Water at the Interface

Water is adsorbed on nonhydroscopic oxides as hydroxyl groups (M—OH) and as molecular water held by H-bonding to the surface hydroxyls. Extensive studies of the adsorption of water on amorphous silica were made by Hair[4] by means of infrared spectroscopy. Most of the physically and chemically adsorbed water can be removed by heating to 800°C in vacuum for a few hours. After this treatment, the infrared spectra show only a low-intensity, narrow band at 3750 cm$^{-1}$. This band is in the region of the stretching vibration for non-H-bonded ("isolated") hydroxyl groups and is attributed to residual surface silanols ($\sim$Si–OH). Rehydration of the silica by water vapor at ambient temperatures produces an increase in intensity of the 3750 cm$^{-1}$ band and the appearance of a new band at about 3650 cm$^{-1}$. These spectral developments correspond to the formation of more isolated surface silanols and silanols close enough to hydrogen bond. Water adsorption on other nonhydroscopic oxides is at least qualitatively similar, that is, there are surface hydroxyls that sometimes interact with each other and are sites for molecular water adsorption.

The surface chemistry of glass departs considerably from that of pure inorganic oxides. All silicate glasses are mixtures of metal oxides dispersed in a silica network. Even in the more water-resistant glass formulations, such as E-glass (16% CaO, 14.5% Al$_2$O$_3$, 9.5% B$_2$O$_3$, 5% MgO, and 55% SiO$_2$), a significant proportion of the nonsilica content is alkali or alkaline earth oxides. These oxides are hydroscopic, so that water adsorption on glass is characterized by the hydration of these oxide microheterogeneities. Pike and Hubbard[5] observed that glass powders adsorbed considerably more water than did a silica powder of comparable surface area.

Not only does glass adsorb relatively thick water films, but this water is strongly alkaline. Aqueous treating baths for fiberglass cloth tend to drift to an alkaline pH and are generally buffered with a weak organic acid to maintain stable bath conditions. The significance of these observations is that the pristine glass surface is capable of adsorbing multilayers of water because of the hydroscopic constituents. After these hydroscopic sites have been washed away, however, the glass surface is essentially a porous silica able to adsorb only a monolayer of water.

According to Fripiat et al.[6] the adsorbed water film on powdered glass is immobile up to a monolayer coverage. The conductivity is low, and they believe that the charge carrier is protionic. Evidently, at these low coverages the water is strongly held by the metal cation and there is less surface diffusional motion than on silica powder having the same amount of adsorbed water. At coverages greater than a monolayer, the surface conductivity of glass increases considerably and charge transfer involves the cation.[7] Studies of glass filaments indicate a high surface conductivity for the pristine filaments, but if they are washed with water before testing the conductivity drops to a level comparable to that of silica filaments.

In discussing corrosion of electronic devices, Cvijanovich[8] considered the immobile layer of hydrogen-bonded water to have negligible effect on corrosion of metal surfaces, but ion movement for corrosion required a cluster of water molecules in addition to the first monolayer of water.

At least a monolayer of water is present on glass and mineral reinforcements during preparation of polymer composites. Even if a composite were prepared with perfectly dry glass, water could penetrate to the interface by diffusion through the polymer. Any imperfections or microcracks in composites formed through stresses set up by differential shrinkage would allow even more rapid entrance of water to the interface. Once water molecules cluster at a composite interface, they are capable of hydrolyzing any bond that conceivably can be formed between resin and glass or other metal oxides. It was demonstrated by Gutfreund and Weber[9] that glass treated with a hydrophobic silylating agent and dried will reabsorb a molecular layer of water upon exposure to the atmosphere. In a study of the water resistance of glass-reinforced thermoset plastics, Vanderbilt[10] found no correlation between water absorption and wet-strength retention of epoxy laminates with various silane coupling agents.

Direct grafting of monomers to minerals during polymerization has been reported.[11] Although such a process can produce intimate contact between polymer and mineral, the ultimate M—O—C bond is readily hydrolyzed and cannot impart exceptional water resistance to the composite.

From the above observations it must be concluded that water cannot be excluded from the interface between resin and a hydrophilic mineral reinforcement and that the effect of water will vary with the nature of the mineral surface. Silane coupling agents do not exclude water from the interface, but somehow function to retain adhesion in the presence of water.

It is fairly well established that silane coupling agents form oxane bonds (M—O—Si) with mineral surfaces where M = Si, Ti, Al, Fe, etc. It is not obvious that such bonds should contribute outstanding water resistance to the interface because oxane bonds between silicon and iron or aluminum, for example, are not resistant to hydrolysis. Yet, mechanical properties of filled polymer castings were improved by addition of appropriate silane coupling agents with a wide range of mineral fillers.[12] Greatest improvement was observed with silica, alumina, glass, silicon carbide, and aluminum needles. A good but somewhat lesser response was observed with talc, wollastonite, iron powder, clay, and hydrated aluminum oxide. Only slight improvement was imparted to asbestine, hydroxyapatite $(Ca_{10}(OH)_2(PO_4)_6)$, titanium dioxide, and zinc oxide. Surfaces that showed little or no apparent response to silane coupling agents included calcium carbonate, graphite, and boron.

Although coupling activity of silanes is not universal to all mineral surfaces, response is broad enough to indicate that silanols need not form water-resistant "oxane" bonds with the surface. Even covalent siloxane bonds are hydrolyzed to silanols by water with an activation energy of 23.6 kcal/mol. Hydrolysis catalyzed by benzoic acid has an activation energy of 6 kcal/mol, which is comparable to the strength of a hydrogen bond. Compression set of silicone rubber has been attributed to a stress relaxation involving hydrolytic breaking and remaking of siloxane bonds of the silicone. Activation energy for stress relaxation of silicone rubber was reported by Osthoff et al.[13] as 22.8 kcal/mol in the absence of catalysts, or 5 kcal/mol in the presence of KOH or benzoic acid.

Andrews et al.[14] studied the permanence of epoxy–glass bonds in 80°C water, with and without added silane coupling agents (Figure 5.1). The initial bond to glass was estimated to be 24 times the expected van der Waals interaction energy, suggesting covalent bonding across the interface. Unmodified epoxy bonds deteriorated rapidly in 80°C water. Failure could be delayed by adding silane coupling agents to the epoxy, but they ultimately failed, indicating hydrolysis of bonds between epoxy and glass.

A significantly different approach was used by Chamberlain et al.[15] to bond a coupling agent chemically to a glass surface. A fluorinated glass surface was allowed to react with a difunctional Grignard reagent to give a firmly bonded organofunctional group. This contrasts with silane coupling agent applied to glass directly from water. Although the coupling agent may have three reactive silanols per molecule, reactive sites on a glass surface are so spaced that not more than one silanol group per molecule can bond to the surface. The remaining silanol groups may condense with

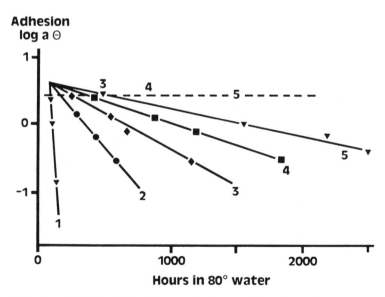

**Figure 5.1.** Pyrex blocks bonded with epoxy adhesive containing silanes of Table 5.1 in water at 80°C. (1) No silane; (2) 0.05% C; (3) 0.05% F; (4) 0.10% C; (5) 1.0% E. The broken line indicates the time when corresponding specimens showed transition from cohesive to interfacial failure [10]. $a$ = Plasticization factor for epoxy saturated with water; $\theta$ = adhesional energy (in J m$^{-2}$).

adjacent silanols to form a siloxane layer or remain partly uncondensed at the surface.

**Glass**

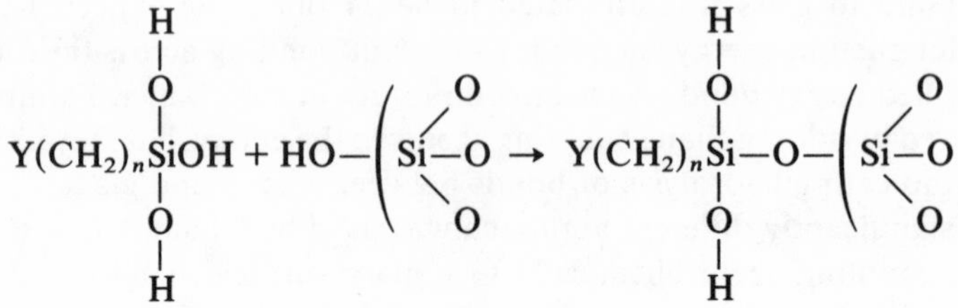

$$\underset{H}{\overset{H}{\underset{|}{\overset{|}{\underset{O}{\overset{O}{|}}}}}} Y(CH_2)_n\overset{\overset{O}{|}}{\underset{\underset{O}{|}}{Si}}OH + HO-\left(\underset{\diagdown O}{\overset{\diagup O}{Si}}-O\right) \rightarrow Y(CH_2)_n\overset{\overset{O}{|}}{\underset{\underset{O}{|}}{Si}}-O-\left(\underset{\diagdown O}{\overset{\diagup O}{Si}}-O\right)$$

**Coupling agent**

$$Y(CH_2)_n MgX + F-\left(\underset{\diagdown O}{\overset{\diagup O}{Si}}-O\right) \rightarrow Y(CH_2)_n-\left(\underset{\diagdown O}{\overset{\diagup O}{Si}}-O\right)$$

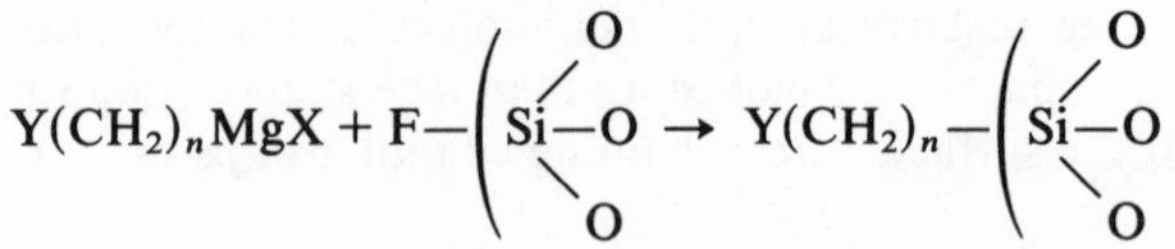

**Chemical bond**

Coupling agents can be removed from glass by hydrolysis of a single siloxane bond, while chemically bonded groups remain bonded to glass until three siloxane bonds are hydrolyzed simultaneously. Chamberlain observed that chemical modification of about 5% of the glass surface gave performance in epoxy laminates comparable to a silane coupling agent applied from water, but that a higher degree of chemical modification gave inferior results. The chemical mobility of coupling agents attached to glass through hydrolyzable bonds appears to be beneficial to composite properties. Any proposed mechanism for coupling through organofunctional silanes must explain how water-resistant composites may be obtained through hydrolyzable bonds across the interface.

## 5.2. Silanol Bonds with Mineral Surfaces

### 5.2.1. General

Silicon is less electronegative than carbon on the Pauling scale.[16] The silicon–carbon bond has 12% ionic character, while the silicon–oxygen bond is 50% ionic in character. Oxane bonds between silicon and most metals are even higher in ionic character.

The bond between silane coupling agents and organic resins may, therefore, be pictured as a bridge of organic covalent bonds, while the bonds between silane coupling agents and mineral surfaces are typically inorganic with a high degree of ionic character.

Covalent bond formation in organic chemistry is determined in a competitive environment by thermodynamics and kinetics of possible competing reactions. Formation and breaking of bonds are not easily reversible. Catalysts may have a strong influence on which bonds will prevail.

Ionic, inorganic chemistry is based more on equilibrium constants than on kinetics. Bonds are formed and broken in reversible reactions determined by concentrations and equilibrium constants. Catalysts may change rates of reaction but are not important in determining final compositions.

Since a hydrolysis-resistant bridge of covalent bonds is readily formed between coupling agent and resin, the water resistance of composites is determined largely by equilibrium conditions of bonds between coupling agent (or resin) and hydroxyl groups of the mineral in the presence of water.

### 5.2.2. Prototype Reactions

In a study of the gas-phase reaction of methyltrimethoxysilane with silica, Hertl[17] reported that the silane reacted with isolated silanol groups on silica to form siloxane bonds. Reactivity of silica was increased by predrying the silica, indicating that reaction was between silanol groups and the methoxysilane and did not involve prehydrolysis of the methoxysilane by surface water.

Dreyfuss[18] studied the reaction of trimethylmethoxysilane and triethylsilanol by gas-liquid chromatography. Immediately after mixing the reagents at room temperature, methanol and the unsymmetrical disiloxane, 1,1,1-triethyl-3,3,3-trimethyldisiloxane, were formed. After about two hours an apparent equilibrium was reached with a major proportion of unsymmetrical disiloxane mixed with the two symmetrical disiloxanes. Addition of acetic acid, n-propylamine, or phenyl-β-naphthylamine as possible catalysts may have increased the rate of redistribution slightly, but did not alter the type or amount of products significantly.

In an extension of this work, Dreyfuss, Fetters, and Gent[19] used poly(butadiene) terminated with dimethylmethoxysilane groups to study the reaction with triethylsilanol. Reaction proceeded readily in benzene at room temperature to form methanol and dimers of the two starting materials. A corresponding condensation reaction between silanated poly(butadiene) and the silanol groups of glass was inferred. Bonding was corroborated by a greater degree of retention of silanated poly(butadiene) on glass slides that were subjected to thorough washing with organic solvents.

When fillers are treated with silane coupling agents by dry-blending or when silanes are added to integral mixes of filler and polymer, it is inferred that alkoxysilanes react directly with surface hydroxyl groups on the mineral and need not prehydrolyze with surface moisture in order to react.

Two silanes were tested as coupling agents in mineral-filled polyester castings by adding 0.3% silane based on the fillers during mixing of filler with resin. Portions of each filler were predried overnight at 100°C and compared with undried filler. Properties of the resulting castings indicate that predrying was especially important with silica filler to obtain full advantage of the silane coupling agent (Table 5.2). Part of the improvement observed with dried filler may be due to improved dispersion in the resin, since water on the particulate surface may prevent complete wet-out of the mineral surface by resin. A cationic vinylbenzyl-functional silane was more uniformly effective with either dried or undried fillers, probably due to its surface-active nature on anionic surfaces (cf. Chapter 6).

Erickson[20] proposed that some water was necessary on fiberglass for coupling agents to function in polyester composites. Glass cloth was treated

**Table 5.2.** Silane Additives in Filled Polyester Castings

| Percent additive (Table 2.2) based on filler | Filler predried overnight 100°C | Viscosity of mix, MPa | Flex strength, of casting, MPa | |
|---|---|---|---|---|
| | | | Dry | 2-hr boil |
| 50% silica (Minusil 5 $\mu$)[a] in P-43[b] polyester | | | | |
| None | yes | 31,800 | 123 | 92 |
| None | no | 20,500 | 107 | 94 |
| 0.3% Silane D | yes | 23,500 | 163 | 143 |
| 0.3% Silane D | no | 23,300 | 138 | 114 |
| 0.3% Silane H | yes | 13,000 | 157 | 140 |
| 0.3% Silane H | no | 22,500 | 145 | 121 |
| 50% Wollastonite (Nyad-400)[c] in P-43 polyester | | | | |
| None | yes | 27,700 | 126 | 97 |
| None | no | 18,000 | 115 | 104 |
| 0.5% Silane D | yes | 20,200 | 145 | 110 |
| 0.5% Silane D | no | 17,400 | 145 | 111 |
| 0.5% Silane H | yes | 20,800 | 138 | 132 |
| 0.5% Silane H | no | 15,900 | 149 | 121 |

[a] Pennsylvania Glass Sand.
[b] Rohm and Haas Co.
[c] Nyco Div. of Processed Minerals, Inc.

with NOL-24 finish (allyltrichlorosilane + resorcinol) in organic solvents after the cloth had been stored at various humidities at 22°C for at least 75 days prior to treatment. Optimum results were obtained with glass stored at about 70% relative humidity, with considerably poorer results at relative humidity below 50%. Modern coupling agents based on organofunctional alkoxysilanes, however, would be expected to differ in reactivity from NOL-24 which is based on a chlorosilane.

### 5.2.3. Equilibrium Conditions at the Interface

It has been demonstrated that silanol groups of silane coupling agents form siloxane bonds with silica or glass surfaces (cf. Chapter 4). These bonds are hydrolyzed during prolonged exposure to water, but re-form when dried. Thus, the reversible nature of siloxane-bonding to these surfaces has been demonstrated, but no direct evidence is available of true equilibrium conditions at the interface. Many observations on silanol reactions, however, suggest true equilibrium conditions.

In order to determine how moisture affects glass and the bond between glass and coupling agent, Vaughan and McPherson[21] exposed heat-cleaned glass and silane-treated woven glass fabric to a relative humidity of 95% at 100°F for a period of 12 weeks. Although heat-cleaned glass suffered some deterioration during exposure, epoxy laminates prepared from exposed silane-treated fabrics were only slightly poorer than comparable laminates from unexposed glass. A series of epoxy prepregs prepared with similar glass fabric was also exposed to a relative humidity of 95% at 100°F for 12 weeks. Laminates prepared from preconditioned prepregs showed no major effect of moisture exposure, except with heat-cleaned glass. Either the bonds of silane coupling agents to glass are not hydrolyzed by moisture in the absence of stress, or else the hydrolyzed bonds are re-formed during the laminating cycle. The silane treatments in some way protected the glass against deterioration under warm, humid conditions.

Evidence for a reversible equilibrium bond between resin and glass is found in the response of glass-reinforced composites to aqueous acids and bases.[22] Both acids and bases are known to be powerful catalysts for hydrolysis (and re-formation) of siloxane bonds. If bonding depended on permanent covalent bonds between resin and glass, mechanical properties of laminates should deteriorate faster in acids or bases than in neutral water. Equilibrium bonding, however, would change in the rate of hydrolysis and condensation with acids or bases, but the point of equilibrium would not be altered. Mechanical properties of laminates in acid or base should, then, be no worse than in neutral water. Portions of typical glass-cloth-epoxy and glass-cloth-polyester laminates were tested for flexural strength after boiling for 24 hours in aqueous solutions at pH 2, 4, 6, 8 and 10. As indicated in Figure 5.2, best retention of properties was in acid or alkaline water. Methacryloxypropyltrimethoxysilane was used on glass for coupling to polyester resins. This silane is known to produce optimum wet-strength retention with polyesters. A hydrophilic, cationic methacrylate coupling agent (Dow Corning® Z-6031) was used on glass for coupling to epoxy resins. This silane would not be expected to produce the most water-resistant composites, but again, deterioration caused by water was the greatest at pH 6.

Pohl and Osterholz[23] demonstrated that condensation and hydrolysis of coupling agent silanols are in true equilibrium in water (cf. Section 3.2.3 and Figure 5.3).

From these data it was possible to estimate the equilibrium constants for hydrolysis of coupling agents on a silica surface (Table 5.3a). Bonding of a typical silane coupling agent $RSi(OH)_3$ to silica has a thousandfold advantage in water resistance over a simple alkoxy bond between a hydroxyl-functional polymer and silica.

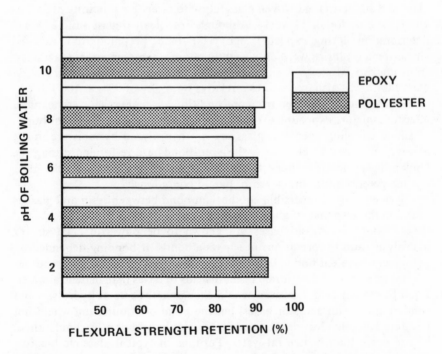

**Figure 5.2.** Strength retention of fiberglass composites after boiling for 24 hours in water at different pHs.

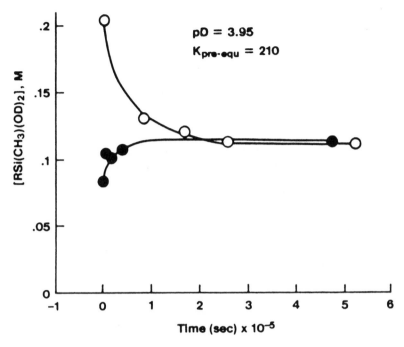

**Figure 5.3.** A plot of the concentration of gamma-glycidoxypropylmethylsilanediol vs. time; $\bigcirc$ = silane diol; $\bullet$ = disiloxane.[23] Courtesy Gordon and Breach Science Publishers.

### 5.2.4. Maintaining Equilibrium Conditions Across the Interface

The equilibrium reaction of major importance in silane-modified mineral-reinforced organic composites is the reaction of oxane bonds with water, where M is the mineral and Si is the coupling agent silicon. Good

$$M-O-Si- + H_2O \rightleftarrows M-O-H + H-O-Si-$$

**Table 5.3a.** Equilibrium Hydrolysis of
Coupling Agents on Silica:
$$\equiv Si-O-SiR_n(OH)_m + H_2O \rightleftarrows \equiv Si-OH + HOSiR_n(OH)_m, \quad n+m=3$$

| Coupling agent on silica | $K_{eq}$ |
|---|---|
| $Si-O-Si(OH)_3$ | $10^{-5}$ (estimate) |
| $Si-O-Si(OH)_2R$ | $10^{-4}$ (estimate) |
| $Si-O-Si(OH)R_2$ | $2.75 \times 10^{-3}$ |
| $Si-O-SiR_3$ | $5 \times 10^{-2}$ |
| $Si-O-R$ | $10^{-1}$ (estimate) |

bonding across the interface (Figure 5.4) requires that the reaction not go too far to the right. Conditions favorable for bonding include

1. a maximum initial formation of MO—Si;
2. a minimum equilibrium concentration of water at the interface;
3. polymer structures that hold silanols at the interface.

**5.2.4.1. Initial Formation of Oxane Linkage.** Silane coupling agents are often applied to minerals from water. Intermediate silanols first hydrogen-bond to the mineral surface but ultimately condense to oxane bonds across the interface. This reaction may be driven to completion by drying at higher temperatures, but under conditions that do not destroy the organofunctional group on silicon.

Fiberglass that had been treated with aqueous methacryloxypropylsilanols was dried under increasingly severe conditions before laminating with polyester resin.[24] Composite properties improved with initial increase in drying temperature, but then dropped off at conditions known to cause loss of methacrylate double bond[25] (Table 5.3b). Vinylsilanes may be heated to higher temperature on glass without detrimental effects to coupling reactivity.

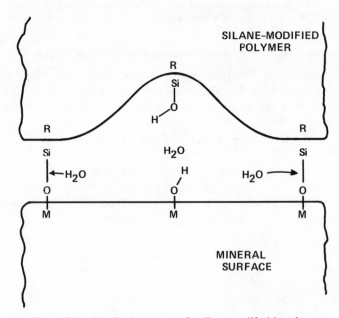

**Figure 5.4.** Idealized structure of a silane-modified interface.

**Table 5.3b.** Polyester Laminates from Glass
Finished with MPS[a]

| Drying time and temp. °C | Flexural strength, MPa | | % Retention |
|---|---|---|---|
| | Dry | 2-hr boil | |
| 24 hr/room | 569 | 526 | 92.4 |
| 7 min/110 | 616 | 583 | 94.6 |
| 15 min/127 | 548 | 545 | 99.4 |
| 7 min/143 | 569 | 554 | 97.4 |
| 7 min/160 | 522 | 481 | 92.1 |

[a] MPS = methacryloxypropylsilanols in water.

Kaas and Kardos[26] reported that condensation of neutral alkoxysilane coupling agents with glass and silica was catalyzed by aliphatic amines. Glass microbeads were treated with silanes from alcohol with and without added amine catalysts and used to prepare polyester composites containing 22% filler by volume. Tensile strengths of the composites were improved markedly by including an amine catalyst in the silane treatment (Table 5.4).

Addition of amines should be considered for all primer applications of silanes in nonaqueous solvents. A mixture of 1% benzyldimethylamine (BMDA) in methacryloxypropyltrimethoxysilane (MPS) in organic solvents is a very effective primer for cross-linkable elastomers used for encapsulation of solar cells.[27]

For treatment of reinforcements from an aqueous silane bath, the silane may be added to a solution of 0.1% amine in water containing an excess of a volatile acid. When a piece of fiberglass cloth was dipped into such a bath containing phenolphthalein indicator, the bath and dipped glass showed no indication of the base, but within 10 sec after dipping, the glass cloth showed a pink color as the volatile acid evaporated. The less volatile

**Table 5.4.** Glass Beads–Polyester Composites

| Surface treatment on glass | Tensile strength of composites | |
|---|---|---|
| | MPa | psi |
| None | 40.6 | 5890 |
| Methacryloxypropylsilane | 54.9 | 7960 |
| Methacryloxypropylsilane + amine | 58.5 | 8490 |
| Vinyltriethoxysilane | 40.5 | 5880 |
| Vinyltriethoxysilane + amine | 50.9 | 7380 |

amine was present only long enough to catalyze silanol condensation and then was lost during complete drying.[28]

Glass microbeads were treated with 0.5% aqueous solutions of methacryloxypropylsilanols in various mixtures of acids and amines. Polyester castings containing 60% by weight of these treated beads were tested for flexural strength after boiling in water. The advantage of alkaline catalysts was best seen in retention of properties after a 72-hr water boil. A nonvolatile amine (aminofunctional silane) was retained at the interface and, although its catalytic activity improved initial composite strength, it caused drastic loss in properties in a 72-hr water boil (Table 5.5).

### 5.2.4.2. Minimum Concentration of Water at the Interface.
Water molecules diffuse through any plastic and thus will reach the interface in mineral-filled composites exposed to a humid environment. Individual water molecules, however, are relatively harmless at the interface unless they are capable of clustering into a liquid phase. The concentration of water at the interface is not determined by the rate of permeation of water through the polymer matrix (silicones and hydrocarbon resins have highest permeability) but the amount of moisture retained at the interface.

Polar functional groups in a resin tend to orient toward a hydrophilic filler's surface. Strong hydrogen bonding across the interface contributes good dry adhesion, but a concentration of polar functional groups also attracts and holds water. The interphase region then becomes plasticized

**Table 5.5.** Composites of 60% Glass Microbeads[a] in Polyester
(0.5% MPS + 0.1% base + excess acid in treatment of beads)

| | | Flexural strength of castings, MPa | | |
| --- | --- | --- | --- | --- |
| Acid in bath | Base in bath | Initial | 2-hr boil | 72-hr boil |
| Control (no silane) | | 117 | 61 | 21 |
| Acetic | None | 138 | 125 | 41 |
| Acetic | BMDA | 144 | 122 | 54 |
| $CO_2$ | $Et_3N$ | 142 | 133 | 55 |
| Acetic | $Et_3N$ | 147 | 119 | 54 |
| Acetic | Aminosilane[b] | 150 | 130 | 32 |
| Using an acrylic-modified epoxy resin-Derekane® 790 | | | | |
| Acetic | None | 109 | 89 | 58 |
| Acetic | BDMA | 115 | 94 | 78 |

[a] = Potters Industries 3000.
[b] = AEAPS Table 2.2.

and weakened by water, so that failure occurs in this interphase region rather than at the true interface or within the bulk of resin.

Vanderbilt[29] correlated the percent strength retention of quartz-filled polymer castings with the effect of boiling water on the polymer itself (Table 5.6). A polyester initially absorbed about 3% water and then hydrolyzed with loss of hydrolytic products on extended contact with hot water. An epoxy resin, with aromatic amine hardener, absorbed an appreciable amount of water, but the linkages within the resin were stable to hydrolysis. Even though chemical bonds were stable to hydrolysis, the epoxy resin lost strength due to the plasticizing action of absorbed water. The hydrocarbon resin absorbed so little water that its strength was virtually unchanged after 7 days in boiling water. A quartz-filled polyester with appropriate silane coupling agent was more resistant to boiling water than an unfilled polyester.

Deterioration of the interface in water is also related to the water resistance of the filler or reinforcement. Quartz generally provides more water-resistant composites than can be obtained with glass. A hydrocarbon (Buton) molding made with sodium sulfate, a water-soluble filler lost over 50% of its strength in 24 hr in boiling water.[29] Ashbee[30] extracted a glass fiber from the center of a polyester composite after prolonged boiling in water and showed the surface of the fiber to be completely pitted. Quartz fibers or glass fibers with good silane coupling agent did not show pitting in comparable tests.

Silane coupling agents may contribute hydrophilic properties to the interface. This is especially true when aminofunctional silanes are used as primers for reactive polymers such as epoxies and urethanes. The primer may supply much more amine functionality than can possibly react with the resin at the interface. Excess unreacted amine at the interface is hydrophilic and is responsible for the poor water-resistance of such bonds. The amount of excess amine at the interface may be minimized by using very dilute solutions of silane in the primer, or by washing the primed surface

**Table 5.6.** Effect of Boiling Water on Quartz-Filled Castings

| Resin | % Increase in Weight[a] | | % Flexural strength retention[b] | |
|---|---|---|---|---|
| | 24 hr | 7 days | 24 hr | 7 days |
| Polyester | 2.9 | −5.9 | 60 | 40 |
| Epoxy | 2.1 | 3.0 | 85 | 80 |
| Buton® (hydrocarbon) | 0.2 | 0.3 | 97 | 96 |

[a] Unfilled resin.
[b] Filled with 50% quartz and appropriate coupling agent.

with water or organic solvent to remove all but a very thin layer of chemically adsorbed silane.[31]

A more effective way to use hydrophilic silanes in primers is to blend than with hydrophobic silanes such as phenyltrimethoxysilane (PS).[32] Only a small proportion of an aminosilane is needed in the mixture to provide adequate reactivity with epoxies or urethanes (Table 5.7). Mixed siloxane primers containing a major proportion of PS also have improved thermal stability typical of aromatic silicones.

**5.2.4.3. Holding Hydrolyzed HOSi Groups at the Interface.** Even though oxane bonds of silane coupling agents to some mineral surfaces are rather easily hydrolyzed, and even though some resins are somewhat hydrophilic, it is possible to obtain water-resistant composites if equilibrium conditions can be maintained between a silane-modified polymer and the mineral surface. Equilibrium conditions are lost if silanol groups resulting from hydrolysis at the interface are physically removed from the interface. Retention of equilibrium conditions is favored by proper control of the morphology of the silane-modified resin at the mineral interface (Figure 5.5).

Practical experience has correlated performance of a composite with resin structure at the interface:

1. Monomers and oils may be removed physically from the interface as individual links to the surface hydrolyze, and thus they are lost for rebonding.

**Table 5.7.** Adhesion of One-Component Urethane to Glass
(cured 5 days in air at room temperature)

| Ratio of silanes I/F in primer | Adhesion to glass (N/cm) | | |
|---|---|---|---|
| | Dry | 2 hr boil $H_2O$ | 5 hr boil $H_2O$ |
| Unprimed control | 3.0 | Nil | Nil |
| 0/100 | c[a] | Nil | Nil |
| 50/50 | c | Nil | Nil |
| 80/20 | c | c | 0.1 |
| 90/10 | c | c | c |
| 95/5 | c | c | c |
| 99/1 | c | c | 1.1 |
| 99.8/02 | c | c | 0.7 |

[a] C = cohesive failure in polymer at $>20$ N/cm.

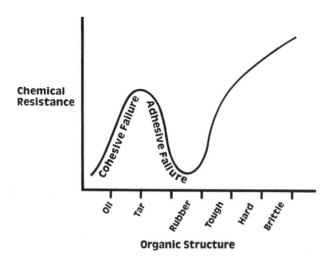

**Figure 5.5.** Relation of polymer morphology of interphase region with chemical resistance of composites.

2. Tarry polymers with viscous flow are self-healing. As silanol bonds are formed by hydrolysis at the interface, they flow to an active site and reform oxane bonds.
3. Flexible and rubbery portions of a polymer retract from the interface as silanol bonds are released through hydrolysis. Water then intervenes and pushes the polymer farther from the surface until complete loss of adhesion results.
4. Silanols formed at a rigid interface cannot move beyond the dimensions of molecule segments, and so are in a position to reform bonds with the original or adjacent active sites on the surface.

Retention of adhesion across the interface is best if the silane-modified polymer is either tarry (has viscous flow) or rigid. It is virtually impossible to obtain a direct water-resistant bond between a flexible, or elastomeric, polymer and a hydrophilic mineral. Water-resistant adhesion may be obtained by modifying the interface through application of a primer or by surface reactions of a silane-modified polymer. The modified interface layer may be very thin (a few hundred angstroms) without altering the flexibility of the polymer itself. In general, thermoplastic elastomers are best bonded through a tarry interface, while vulcanized elastomers are best bonded through a cross-linked resinous interface.

The effect of polymer morphology on adhesion was illustrated in a study of acrylic copolymers with constant silane modification but varying glass transition temperatures.[33] A series of copolymers of methyl methacrylate (MMA) and ethyl acrylate (EA) was prepared, having glass transition temperatures between those of the homopolymers, MMA = 105°C and EA = −23°C. Molecular weights were maintained at about 10,000 by including 1.5% mercaptopropyltrimethoxysilane as chain transfer agent. Additional silane modification was contributed by 1% methacryloxypropyltrimethoxysilane in each composition.

Films on glass microscope slides were dried for 30 min at 100°C and examined for adhesion both dry and after soaking for one day in water. All films showed good dry adhesion, but copolymer films of intermediate composition that were flexible or rubbery showed poorest wet adhesion (Table 5.8). Similar acrylic copolymers without silane modification generally had very poor wet adhesion to glass except for soft tacky films.

Good examples of hydrolytically stable cross-linked structures are silica and silicate rocks. Although every oxane bond in these structures is hydrolyzable a silicate rock is quite resistant to water. Each silicon is bonded to four oxygens under equilibrium conditions with a favorable equilibrium constant for bond retention. The probability that all four bonds to silicon can hydrolyze simultaneously to release soluble silicic acid is extremely remote. With sensitive enough analytical techniques it is possible to identify soluble silica as it leaches from rocks, but an individual rock will survive in water for thousands of years.[34]

Silane coupling agents provide oxane bonds between organic adhesives and metals or glass, but the interphase region is not highly cross-linked. Even though silane coupling agents are trifunctional in silanol groups there is a strong tendency for the silanols to condense to cyclic oligomers rather than to cross-linked structures. Addition of certain polyalkoxyfunctional silanes to standard coupling agents provides a high degree of siloxane

**Table 5.8.** Methyl Methacrylate (MMA), Ethyl Acrylate (EA) Copolymers

| Copolymer composition | A | B | C | D | E | F |
|---|---|---|---|---|---|---|
| MMA | 100 | 70 | 50 | 40 | 30 | 20 |
| EA | — | 30 | 50 | 60 | 70 | 80 |
| Mercaptosilane | 1.5 | 1.5 | 1.5 | 1.5 | 1.5 | 1.5 |
| Methacrylate silane | 1.0 | 1.0 | 1.0 | 1.0 | 1.0 | 1.0 |
| Film adhesion[a] to glass | Very hard | Hard | Tough | Rubber | Soft | Tarry |
| Dry | 10 | 10 | 10 | 10 | 10 | 10 |
| Wet (1 day) | 10 | 10 | 9 | 5 | 8 | 9 |

[a] Rate of adhesion: 0 = nil to 10 = excellent.

cross-linking, which gives water-resistant bonds that should be stable for a thousand years.

The preferred polyfunctional silane cross-linker is 1,2-bis(trimethoxysilyl)ethane $(CH_3O)_3SiCH_2CH_2Si(OCH_3)_3$. This material was tested in mixtures with commercial silane coupling agents in several systems.[35]

Primers were evaluated by forming thin films of polymer (by fusion or by curing) on primed microscope slides or metal coupons and soaking in water to determine time to failure (Table 5.9). In this test, water diffuses almost instantly to the interface, and measurement of water resistance of the bond is virtually independent of the rate of diffusion. Time to failure is shortened more than a thousandfold over results indicated on tests with lap shear joints.

In summary, the bond between a silane-modified resin and a hydrophilic mineral surface is pictured as existing under equilibrium conditions of oxane bond formation and hydrolysis to silanols in the presence of water. Most favorable bonding in the presence of water is obtained with a hydrophobic interphase region that is highly cross-linked. Hydrolytic action of water at the interface may actually benefit total composite performance by allowing a chemical relaxation of stresses across the interface through equilibrium bonding. Although accelerated tests show that bonds of polymers to silane-primed glass and metals are not impervious to water, it is possible to design systems in which the bonds rival the stability of silicate rocks in water and should be able to survive under ambient conditions for a thousand years.

From a chemical standpoint, the best water resistance results if the interface region adjacent to glass or filler is highly cross-linked. From a mechanical viewpoint, optimum bonding requires a flexible interphase region between a mineral filler and a rigid polymer. Eckstein deposited very

**Table 5.9.** Adhesion of Cross-Linkable EVA to Stainless Steel

| Primer on stainless steel (304) | 90° peel after days in 70°C water (N/cm) | | | |
|---|---|---|---|---|
| | 1 | 4 | 10 | 20 |
| MPS | 7.7 | Nil | — | — |
| 9/1 MPS/X-linker[a] | 15.4 | 5.0 | Nil | — |
| 1/1 MPS/X-linker | c[b] | 11.6 | Nil | — |
| 1/3 MPS/X-linker | c | c | c | 16 |
| X-linker alone | c | 23 | 23 | 24 |

[a] X-linker = $(CH_3O)_3SiCH_2CH_2Si(OCH_3)_3$.
[b] c = cohesive failure in the polymer at over 30 N/cm.

thin layers of polyester (200–1600 Å) on glass braids and tested the composites with a DuPont Dynamic Mechanical Analyzer.[36] Gamma MPS on the glass interdiffused with the polyester so that thin films had the character of the interphase region of standard fiberglass composites. Highest shear strength resulted when the silanepolyester was soft and ductile ($T_g$ and modulus lower than the bulk resin). Tryson and Kardos noted a similar improvement in epoxy composites when a thin ductile layer was formed between the glass and epoxy matrix.[37] It is interesting that Jang et. al.[38] showed that mixtures of MPS and the cross-linker L deposited on glass from water with preferential adsorption of the cross-linker next to the glass. This may lead to the ideal combination of chemical stability and mechanical strength of bond. The interface of coupling agent and glass should be highly cross-linked with siloxane bonds, but the interphase region extending several hundred angstroms from the glass will be a tough polymer resulting from interdiffusion of coupling agent with matrix polymer.

## 5.3. Bonding to Metals

### 5.3.1. General Concepts

There are many individual examples of the use of silane coupling agents to improve the water resistance of polymer bonds to metals. Silanes are generally useful, but they differ widely with different metals. There has been little published data on systematic studies of comparisons of silane adhesion promoters with the wide range of structural metals to be bonded. An effort is made here to accumulate such data and arrive at a general understanding of silane adhesion to metals.

Surfaces of common metals under ambient conditions are hydrated oxide surfaces. Zettlemoyer estimated that glasses, silicas and clays have about one silanol group per 50–100$\text{Å}^2$, and that a similar population of hydroxyls exists on aluminum, iron and other metal oxides.[39]

Bolger[40] studied the adhesion of hydroxyl and carboxyl-modified polymers to metal surfaces and correlated adhesion to the probabilities of acid base interaction across the interface derived from the IEPS of the oxide and the $pK_a$ of the polar groups in the polymer. He found the adhesion of functional polymers was best when the difference between IEPS and pK was other than maximum or minimum, but of intermediate value.

Fowkes[41] demonstrated that acid–base reaction across the interface was beneficial to adhesion. The addition of appropriate acid or basic solvents could often overcome a mismatch of acid–base attraction across the interface.

Boerio and co-workers[42,43] studied the effect of silane primers in bonding steel, titanium, and aluminum with epoxy adhesives. Silane primers improved the water resistance of bonds in general, and were dependent upon pH of application to iron, but not to titanium and aluminum. One of the beneficial effects on aluminum was that the silane primers improved the hydrothermal stability of the oxide. Walker[44] showed that silane coupling agents were effective in improving the water resistance of bonds of epoxy and urethane paints to aluminum, steel, cadmium, copper, and zinc (see Tables 6.23 and 6.24).

Silane coupling agents that provide water-resistant bonds of polymers to silica and glass are generally effective in improving adhesion to metals, but the degree of improvement varies widely with different metals. The siloxane bonds between coupling agents and silica in water are in true equilibrium of hydrolysis and condensation with very favorable equilibrium constants for bond formation (Table 5.3a).

There is no firm evidence that silane coupling agents have a similar mechanism of metal-oxane bonding ($M-O-Si$) with metal hydroxide surfaces, but it seems most probable because almost all metals are found in nature as silicate minerals. It would be expected that equilibrium constants for hydrolysis of various ($M-O-Si$) bonds would differ, but the factors that determine such constants are not known.

In a continuation of our study[45] on the bonding of polymers to metals, thin films of carboxy-modified polyolefins were fused against primed and unprimed metal surfaces and soaked in water until the 90° peel strength dropped below 15 N/cm (10 lb./in.)[45] Water diffuses almost instantly to the interface under these conditions which provides an enormous acceleration of failure as compared to lap shear tests where moisture must diffuse in from the edges. Attempts were made to correlate water resistance of bonds to metals with fundamental properties of metal hydroxides that might be expected to effect bond stability in water.

### 5.3.2. Factors in Adhesion to Metal

**5.3.2.1. Isoelectric Point of the Oxide.** Parks[46] collected data on isoelectric points (IEP) of many oxide surfaces. The IEP of a surface is defined as the pH in water at which the zeta potential of the surface is zero. It is a good measure of the relative acid–base nature of the surface.

**5.3.2.2. Ionic Character of the Oxide Bond.** Pauling[47] has calculated the percent of ionic character of typical metal oxide bonds from the difference in electronegativity between the metal and oxygen. There is no direct relationship between percent ionic character and the IEP of the surface or the solubility of the hydroxide.

**5.3.2.3.  Solubility of the Metal Hydroxide in Water.** Solubility of metal hydroxides in water as a function of pH were plotted by Kragten[48] as shown in Figures 5.6. If solubility of the metal hydroxide were a major factor in adhesion, it would be expected that adhesion on many metals would be a function of pH of the aqueous environment. Metals whose oxides are soluble in acids (Mg, Ni, Cr, Pb, and Zn) might be expected to form bonds susceptible to debonding by acids. Metals of amphoteric hydroxide (Al, Zn) might be susceptible to debonding by both acids and bases. The extreme insolubility of Sn IV oxide over a wide pH range suggests that bonds to tin should have outstanding water resistance over a wide pH range.

**5.3.2.4.  Chelation of Metals by Ethylenediamine.** The stabilities of metal chelates with ethylenediamine were described by Kragten[48] as a function of pH. If such chelation is an important factor in bonding an ethylenediamine-functional silane (Z-6020) to metals, it would be expected that this silane would show unique properties in bonding to copper and nickel as compared to aluminum and silicon. Bonding with chelating metals would also be a function of pH.

**5.3.2.5.  Other Factors.** Metal oxide surfaces may also differ in other factors that effect adhesion through silane coupling agents. Spacing of hydroxyl groups on silicon match the spacing of comparable groups on silane coupling agent oligomers. Hydroxyls of metal oxide surfaces may have varying degrees of mismatching of spacing for optimum bonding.

Metal oxide surfaces may differ in their ease of contamination by polar organic molecules in the environment. It appears, that all clean metal surfaces have such high surface energy that they are contaminated rapidly in ordinary atmospheric environment.

### 5.3.3.  Early Tests on Adhesion

Earlier work[45] with three types of polymers showed little correlation of adhesion (water resistance) with the properties of metal oxides. These systems included:

1. A PVC plastisol containing 0.5% Q1-6012 (partially prehydrolyzed 3-(2-aminoethyl) aminopropyl trimethoxysilane) (AEAPS).[49]
2. A cross-linkable ethylene-vinylacetate copolymer with gamma-methacryloxy propyltrimethoxy silane (MPS) primer.[50]

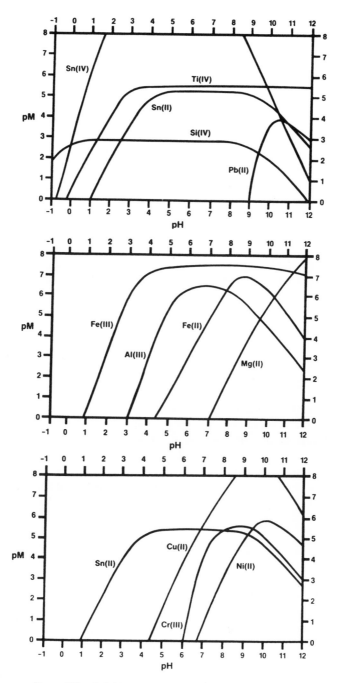

**Figure 5.6.** Solubilities of metal ions as a function of pH.

3. Copolymers of styrene with 4 mol% methacrylic acid, dimethyl-aminoethylmethacrylate, or MPS.[51]

Adhesion was rated qualitively by attempting to loosen films from the surface with a razor blade. Samples were soaked in water at room temperature and examined periodically for adhesion to determine whether the film could be loosened. Films were rated according to the length of time they withstood the action of water (Table 5.10).

Adhesion of PVC plastisol with 1% added Q1-6012 was repeated in quadruplicate samples on clean vapor deposited metals.[52] Adhesion was measured after fusion to six metals for 5 min at 170°C, ageing for 7 days at room temperature and during soaking in water at pH 4,7, and 10 for up to 35 days. (Table 5.11).

Combining the data of Tables 5.10 and 5.11 indicate the following.

1. Silane primers are exceptionally effective on silicon and glass, but also good with titanium, tin, aluminum, and brass. The isoelectric point of the oxide surface does not appear to be a critical factor in adhesion.
2. Acid-, and base-modified polymers show a mild trend of improved bonding to high-IEP and low-IEP surfaces, respectively, showing some acid–base interraction in adhesion.
3. Solubility of the oxide in water at the test pH did not seem to be a factor in adhesion. As long as there was adhesion of polymer to a surface, there was no attack of the metal under the polymer film, even though the bare metal was attacked by the solution (e.g., Mg, Ni, and Zn at pH 4).
4. Chelate formation between the silane and metal surface (e.g., AEAPS on Ni at pH 10) was very detrimental to adhesion.

### 5.3.4.  Bonding Carboxyl-Modified Polyolefins

A new series of tests was conducted on adhesion of carboxyl-modified polyolefins to unprimed and primed metal surfaces.[52] Modified polyolefins studied were:

Kraton 1901X, a maleic anhydride-modified styrene-butadiene block copolymer (Shell Chemical Co.)
Polybond 1016, an 80/20 propylene/ethylene copolymer modified with 6% acrylic acid (BP polymers)
Primacor 5980, a polyethylene modified with 20% acrylic acid (Dow Chemical Co.)

**Table 5.10.** Adhesion[a] of Polymers to Metals Through Functional Groups

| Metal surface | Nature of oxide surfaces | | | Adhesion of Modified Polymers (water resistance) | | | | |
| --- | --- | --- | --- | --- | --- | --- | --- | --- |
| | IEPS of oxide | % Ionic M—O | pM (aq) pH = 7 | PVC plastisol Q1–6012 | X-linkable EVA gamma-MPS | Styrene copolymers (4 mol% modified) | | |
| | | | | | | —COOH | —N(CH$_3$)$_2$ | —Si(OCH$_3$)$_3$ |
| Si | 2 | 44 | 3.0 | 5 | 5 | 1 | 2 | 5 |
| Glass | 4-5 | 50 | | 5 | 5 | 1 | 2 | 5 |
| Sn | 6 | 43 | 8.0 | 3 | 4 | 1 | 3 | 4 |
| Ti | 6 | 58 | 5.5 | 5 | 1 | 1 | 2 | 4 |
| Cr | 7 | 53 | 4.8 | 2 | 1 | 1 | 2 | 2 |
| Al | 9 | 55 | 6.5 | 4 | 5 | 5 | 5 | 4 |
| Zn | 9 | 53 | 1.4 | 1 | 1 | 2 | 2 | 5 |
| Fe | 9-11 | 46 | 7.5 | 3 | 1 | 2 | 1 | 4 |
| Brass | 9 | 44 | 5.5 | 4 | 3 | 4 | 2 | 4 |
| Ni | 11 | 44 | 0.5 | 1 | 1 | 3 | 1 | 1 |
| Mg | 12 | 79 | 0 | 1 | 2 | 3 | 1 | 1 |

[a] 0 = Poor dry adhesion; 1 = Lost adhesion in 4 hr; 2 = Good 4 hr, poor 1 day; 3 = Good 1 day, poor 4 days; 4 = Good 4 days, poor 7 days; 5 = Good >7 days.

**Table 5.11.** Retention of Adhesion of PVC Plastisols to Metals at
Various pH's

| Metal (IEPS) | pM of oxide | pH of Test | 180° Peel (N/cm) | | | | |
|---|---|---|---|---|---|---|---|
| | | | Dry Peel | Wet Peel after soaking | | | |
| | | | | 1–3 Days | 9 Days | 16 Days | 35 Days |
| Si(2) | 3 | 4 | 11.2 | 5.4 | 7.5 | 12.1 | 14.7 |
| | 3 | 7 | | 10.7 | 10.5 | 13.8 | 13.5 |
| | 2 | 10 | | 11.5 | 10.7 | 13.0 | 13.1 |
| Ti(6) | 5.5 | 4 | 13.9 | 9.3 | 11.5 | 10.7 | 13.1 |
| | 5.5 | 7 | | 14.5 | 12.6 | 7.0 | 11.5 |
| | 5.5 | 10 | | 13.3 | 11.7 | 14.3 | 11.5 |
| Al(9) | 3.0 | 4 | 8.7 | 11.9 | 11.7 | 10.9 | 11.5 |
| | 6.5 | 7 | | 11.7 | 11.7 | 11.2 | 12.1 |
| | 4.5 | 10 | | 8.6 | 10.2 | 10.8 | 10.5 |
| Ni(11) | Sol. | 4 | 14.0 | 3.5 | 5.2 | 6.3 | 8.4 |
| | 0.5 | 7 | | 1.7 | 1.8 | 3.2 | 7.5 |
| | 6.0 | 10 | | 1.9 | 1.9 | 1.8 | Nil |
| Mg(12) | Sol. | 4 | 6.5 | 7.7 | 8.6 | 8.7 | 12.2 |
| | Sol. | 7 | | 9.3 | 9.6 | 14.0 | c |
| | 5.5 | 10 | | 7.4 | 12.3 | 15.2 | c |

c = cohesive failure in polymer at over 20 N/cm.

Primacor 3960, a polyethylene modified with 9.7% acrylic acid (Dow
   Chemical Co.)

Primacor 3960 was also bonded to metal primed with 1% aqueous AEAPS
at pH 4 and 10, and with a Zn-modified acid-functional silane to provide
an ionomer bond between polymer and primer.[53]
   Surfaces for bonding were:

| | |
|---|---|
| Si | Hyperpure silicon wafers cut from single crystals |
| A-Glass | Fisher brand precleaned soda-lime glass microscope slides |
| E-Glass | Flat slabs of E-glass supplied by a fiberglass manufacturer |
| Tin-plated steel | Unknown |
| Titanium panels | Timet HTK-6812 |
| Zn-galvanized steel | Unknown |

| Cr-plated steel | Unknown |
| Ni-plated steel | Unknown |
| Cold-rolled steel | SAE 1010 |
| Stainless steel | Type 304 |
| Aluminum | 303 H-14 mill finish |
| Aluminum/copper alloy | 2024 T3 |
| Magnesium | Dow Ingot A291D |
| Brass shim stock | 69.7% Cu, 30.3% Sn |
| Bronze | U.S. Cent. |
| Copper | Copper-clad epoxy laminate |
| Silver | Silver-plated copper foil |
| Gold | Firmilay type III, hard |
| Lead | Solder-coated brass sheet |

Metal coupons were scrubbed with Comet® cleanser, washed with water, and rinsed with acetone. Coupons were considered to be clean if water spread on the surface. Primers based on 1% aqueous AEAPS at pH 4 and 10 were wiped on the surfaces and air-dried for 30 min at room temperature. The Zn-modified carboxy-modified silane was applied at pH 4, and dried 15 min at 175°C.

Polymers were fused to the surfaces in a Carver press with platens at 220°C under light pressure to produce films about 0.5–1.0 mm thick. Coated coupons were soaked in water at room temperature at pH 7 (one series at pH 4) and examined periodically for adhesion. Time to failure was measured as the time when the 90° peel strength decreased to less than 15 N/cm (10 lb./in.). Those that showed no failure in 7 days soak at room temperature were tested further in boiling water. The ratings of Table 5.10 were extended to two more units.

The adhesion of four carboxy-modified polyolefins to metals is shown in Table 5.12 relating some of the properties of the oxides with the water resistance of bonds with the polymers. Maleic acid groups were much less effective than those derived from acrylic acid copolymers. Among acrylic-modified polymers, 20% acid modification was generally poorer than 6% or 9.7% modification. Adhesion showed no consistent trend with the four properties of oxides listed. Adhesion was very poor to surfaces with low (silica glass) or high (Mg, Ag, A$\mu$) IEPS. In between, there was very good adhesion to Sn, Cr, Al, stainless steel, brass, and copper, but poorer adhesion to Ti, Zn, solder, and Ni.

Ionic character of the oxide was not consistent with adhesion. Both high (Mg) and low (solder) ionic character gave poor bonds while those in-between varied from very poor to very good.

**Table 5.12.** Adhesion of Carboxy-Modified Polyolefins to Metals[a]

| Metal surface | Nature of oxide surfaces | | | Adhesion of polymers (water resistance) | | | |
|---|---|---|---|---|---|---|---|
| | IEPS of oxide | % Ionic M—O | pM (aq) pH = 7 | SBR[b] block maleic | 80/20[c] PP/PE 6% acid | Primacor® 5980 PE 20% acid | Primacor® 3960 PE 9.7% acid |
| Si | 2 | 44 | 3.0 | 1 | 0 | 1 | 2 |
| Glass | 4–5 | 50 | — | 0 | 1 | 1 | 2 |
| Sn | 4–5 | 43 | 8.0 | 1 | 6+ | 6 | 5 |
| Ti | 6 | 58 | 5.5 | 1 | 1 | 1 | 2 |
| Cr | 7 | 53 | 4.8 | 1 | 5 | 2 | 6+ |
| Al | 9 | 55 | 6.5 | 1 | 6 | 1 | 5 |
| Al-Cu | 10 | — | — | 3 | 4 | 2 | 6+ |
| Zn | 9 | 53 | 1.4 | 1 | 2 | 3 | 4 |
| Solder | 10 | 33 | — | 1 | — | 3 | 1 |
| Cold-rolled steel | 10 | 46 | 2.0 | 2 | 3 | 2 | 3 |
| Stainless steel | 11 | | — | 3 | 6+ | 2 | 6+ |
| Brass | 10 | | — | 1 | 1 | 5 | 5 |
| Cu | 11 | 44 | 5.5 | 1 | 6+ | 5 | 6+ |
| Ni | 11 | 44 | 0.5 | 3 | 3 | 1 | 4 |
| Mg | 12 | 79 | 0 | 2 | 3 | 3 | 3 |
| Ag | 12 | — | 1.0 | 1 | 1 | 1 | 2 |
| Au | — | — | — | 0 | 1 | 0 | 0 |

[a] Rating: 0 = poor dry adhesion; 1 = good dry, poor 4 hr in water; 2 = good 4 hr, poor 1 day; 3 = good 1 day, poor 4 days; 4 = good 4 days, poor 7 days; 5 = good 7 days, poor 4 hr additional boil; 6 = good 4 hr, poor 12 hr boil; 6+ = no failure in 7 days r.t. + 12 hr boil.
[b] Kraton 1901-X.
[c] Polybond 1016.

Solubility of the oxide at pH 7 did not correlate with the water resistance of bonds. Oxides of copper and titanium have similar solubilities at pH 7, yet adhesion to copper was very good, while adhesion to titanium was poor.

Primacor 3960, a polyethylene modified with 9.7% acrylic acid was then fused to metal surfaces with different silane primers and examined for water resistance (Table 5.13). The most obvious effect of silane primers is that they all improve the adhesion of acid-modified polymer to silicon and glass, and to a lesser degree adhesion to Ti, solder, cold-rolled steel, silver, and gold. Adhesion was good to all metals except magnesium and gold. E-glass and soda-lime glass gave very similar adhesion with silane primers.

The pH of the AEAPS primer was of little significance on many surfaces and varied on the others. An alkaline solution (pH 10) was best on Zn, Ni, and Aμ, while an acid solution (pH 4) was best on Ti, Al, Zn, Ag, cold-rolled steel, and Cu. It was surprising that the pH of the water soak was of very little significance in most cases. Acified water (pH 4) was a more severe environment than neutral water (pH 7) only with Ti, Al, solder, stainless

**Table 5.13.** Adhesion of Primacor® 3960 to Silane-Primed Metals[a]

| Metal surface | IEPS of oxide | % Ionic M—O | pM (aq) pH = 7 | Adhesion of Polymers (water resistance) | | | | |
|---|---|---|---|---|---|---|---|---|
| | | | | No primer | 1% AEAPS pH 10 | 1% AEAPS pH 4 | 1% AEAPS pH 4[a] | Zn Ionomer to silane |
| Si | 2 | 44 | 3.0 | 2 | 6+ | 6+ | 6+ | 6 |
| Glass | 4-5 | 50 | — | 2 | 6+ | 6+ | 6+ | 5 |
| Sn | 4-5 | 43 | 8.0 | 5 | 5 | 5 | 5 | 4 |
| Ti | 6 | 58 | 5.5 | 2 | 3 | 5 | 3 | 2 |
| Cr | 7 | 53 | 4.8 | 6+ | 6+ | 5 | 6+ | 3 |
| Al | 9 | 55 | 6.5 | 5 | 2 | 6+ | 5 | 3 |
| Al-Cu | 10 | — | — | 6+ | 6+ | 6+ | 6+ | 3 |
| Zn | 9 | 53 | 1.4 | 4 | 5 | 4 | 4 | 1 |
| Solder | 10 | 33 | — | 1 | 3 | 6 | 2 | 0 |
| Cold-rolled steel | 10 | 46 | 2.0 | 3 | 6 | 5 | 5 | 4 |
| Stainless steel | 11 | — | — | 6+ | 5 | 6+ | 6 | 4 |
| Brass | 10 | — | — | 5 | 5 | 5 | 5 | 6 |
| Cu | 11 | 44 | 5.5 | 5 | 6 | 6+ | 5 | 6+ |
| Ni | 11 | 44 | 0.5 | 4 | 5 | 4 | 3 | 4 |
| Mg | 12 | 79 | 0 | 3 | 2 | 2 | 2 | 2 |
| Ag | 12 | — | 1.0 | 2 | 2 | 5 | 3 | 1 |
| Au | — | — | — | 0 | 3 | 1 | 1 | 0 |

[a] Rating: 0 = poor dry adhesion; 1 = good dry, poor 4 hr in water; 2 = good 4 hr, poor 1 day; 3 = good 1 day, poor 4 days; 4 = good 4 days, poor 7 days; 5 = good 7 days, poor 4 hr additional boil; 6 = good 4 hr boil; 6+ = no failure in 7 days r.t. + 12 hr boil.
[b] Soak in water at pH 4, all others at pH 7.

steel, Cu, and Ni. Neutral water seemed to more severe on Cr and Ag. The acid soak attacked the metals, Mg, Zn-galvanized steel, and plated tin where the metal was bare, but the metal was protected as long as the polymer retained adhesion. This suggests that corrosion of polymer-coated metals must generally be preceeded by loss of adhesion of the coating. Rate of corrosion is more related to the adhesion of the coating than to the reactivity of the metal with the corrosive environment.

The ionomer bond to silane primer was not optimized in the ratio of zinc to carboxyl group, but in general it paralled the performance of AEAPS (Z6020) primers on metals.

### 5.3.5. Conclusion

There is no simple correlation between adhesion of polymers to various metals with any one fundamental characteristic of the oxide surface. Certain carboxy-modified polymers developed very water-resistant bonds to some metals, but not to others. Bonding of polymers to metals through silane coupling agents was much more universal, and more effective, than simple acid–base character would have predicted. Although there is no firm evidence that bonding of silane primers to metals involves a true equilibrium of hydrolysis and condensation in the presence of water, data on adhesion suggests such an equilibrium condition. Each metal has its own equilibrium constant of adhesion to silanols in the presence of water, and individual values of the constants are not known. The concept of equilibrium bonding to metals, however, suggests methods of improving adhesion of silanols that have proved effective on glass and silica (cf. Section 5.2.4).

- Drive the original condensation reaction to completion.
- Develop hydrophobic interface regions to reduce the amount of water at the interface under equilibrium conditions. This is more important than rate of diffusion of water to the interface.
- The interphase region should be highly cross-linked in the finished composite. Adhesion may be improved by adding cross-linking silanes to the primer.

Formulation of appropriate silanes into primers for individual polymer–metal adhesive bonds should be capable of eliminating water-induced failure under ambient conditions as one of the variables in choosing adhesive systems.

### 5.4. Silane Bonds with Polymers
#### 5.4.1. General

Although simplified representations of coupling through organo-functional silanes often show a well-aligned monolayer of silane forming a covalent bridge between polymer and filler, the actual picture is much more complex. Coverage by hydrolyzed silane is more likely to be equivalent to several monolayers. The hydrolyzed silane condenses to oligomeric siloxanols that initially are soluble and fusible, but ultimately can condense to rigid cross-linked structures. Contact of a treated surface with polymer matrix is made while the siloxanols still have some degree of solubility. Bonding with the matrix resin, then, can take several forms.

The oligomeric siloxanol layer may be compatible in the liquid matrix resin and form a true copolymer during resin cure. It is also possible to have partial solution compatibility in which case an interpenetrating polymer network forms as the siloxanols and matrix resin cure separately with a limited amount of copolymerization. Probably all thermosetting resins are coupled to silane-treated fillers by some modification of these two extremes.

Interdiffusion of siloxanol segments with matrix molecules may become a factor in bonding thermoplastic polymers. This must be the case when a silane–thermoplastic copolymer is used as primer or coupling agent for the corresponding unmodified thermoplastic (5.4.3.2). A siloxanol layer may also diffuse into a nonreactive thermoplastic layer and then cross-link at fabrication temperature. Structures in which only one of the interpenetrating phases cross-links have been designated pseudointerpenetrating networks (pseudo-IPN).[54] Amine-functional silanes (E, F, and H of Table 2.2) probably function in this manner in coupling to polyolefins and possibly to other thermoplastics. Performance is often improved by adding a high temperature peroxide to the coupling agent to aid in cross-linking the siloxanol oligomers. Layers of amine-functional siloxanols, in the absence of matrix resins, cure at 150°C to very hard, tough films.

### 5.4.2. Thermosetting Resins

The most important application of mineral fillers and reinforcements in polymer composites is with thermosetting resins. Silane coupling agents were originally designed for these resins and their performance is best understood in this application.

The organofunctional group of a silane coupling agent is selected for chemical reactivity with the resin during cure. Performance of composites is generally directly related to the degree of reactivity of the organofunctional group with the resin. A comparison of fiberglass–polyester laminates prepared with 12 different unsaturated silanes on the glass showed that wet strengths (after a 2-hour boil) followed the relative reactivity of the silane with styrene (Table 1.1). Copolymerization with maleate and fumarate structures in the polyester may also be a factor in the relative reactivity of the silanes.

There is no correlation between polarity of the organofunctional group on the silane coupling agent, or wettability of silane-treated glass, and the effectiveness of the silane coupling agent in polyester composites. Glass treated with chloropropylsilane coupling agent has a relatively high surface energy and is readily wet by resin solution. Such treatment, however, is completely ineffective in laminates with polyester, melamine, or phenolic resin. It is very effective with epoxies where chemical reaction is possible between the chloropropyl group and the epoxy curing agent.

Critical surface tensions of silane-treated surfaces derived from contact angle measurements were shown to vary with given silanes according to methods of application and drying (4.5.3). A better measure of compatibility of silane treatments with a resin than critical surface tensions is to compare the solubility parameters ($\delta$) of R–H compounds corresponding to R-groups of silane coupling agents.

Polyester laminates were prepared from glass cloth treated with silanes having saturated and unsaturated R-functional groups on silicon. Wet strengths (after a 2-hour boil) show no particular correlation with variations in $\delta$ of the coupling agent (Figure 1.6).

Among nonreactive groups like ethyl, chloropropyl, dichlorophenyl, cyanopropyl, and hydroxypropyl it was found that none of the treatments was much of an improvement over bare glass. Most of them were substantially poorer than the control. Properties of similar polyester laminates with reactive silanes show poor correlation with polarity, but good correlation between performance and reactivity of the silane in free-radical polymerization. A comparison of a methacrylate ester-functional silane and a crotonate ester silane with similar $\delta$ values of about nine show this difference to be related to rectivity. The methacrylate is among the best known silane coupling agent for polyesters, while the crotonate is among the poorest of unsaturated-functional silanes in this application. Reactivity of the silane in copolymerization is obviously of much greater significance than polarity or wettability of the treated glass surface. Wetting of a reinforcement by resin may be of practical importance in obtaining good dispersion of filler in resin and in obtaining void-free laminates with glass fibers, but it is not a primary factor in developing water-resistant bonds across the interface.

Interpenetrating-polymer-network structure was postulated by Lee and Craig[55] in adhesion of polyimide coatings to silane primers on silicon wafers. Primers based on aminopropyl silanes provided good initial adhesion across the interface. Cured films of polyimide on primed silicon were exposed to air at 400°C until the aminopropyl groups of the silane were completely lost, but good adhesion of the polyimide film was retained. When the primer film was preheated at 400°C to burn off aminopropyl groups, a polyimide film had poor adhesion to the treated surface.

It was postulated that aminopropyl groups of the primer had initial compatibility with amic acid precursors of the polyimide film, and may have actually formed some silane-functional amides. During cure, the cyclic imide structure was preferred, forcing the aminopropyl siloxane into a separate phase where it condensed as a polysiloxane interpenetrating the polyimide structure. High-temperature oxidation burned off the aliphatic amine groups, leaving silica as the interpenetrating phase. A preoxidized primer layer was essentially a network of silica that had no mechanism for interpenetrating the polyimide coating.

Primers for silicone rubbers and resins are essentially mixtures of orthosilicate and orthotitanate esters. The partially hydrolyzed primer layer has some compatibility with the silicone, but polymerizes separately by hydrolysis and condensation to mixed silica–titania networks. Such inorganic networks have essentially unlimited heat stability and allow retention of adhesion at the highest temperatures of silicone or polyimide coatings.

### 5.4.3. Thermoplastic Resins

**5.4.3.1. Bonding Through Solution Compatibility or Diffusion.** Thermosetting resins require reactive silanes for coupling to glass. Among reactive silanes there was no clear correlation between solution compatibility of the silane and laminate properties. Thermoplastics may also be coupled to glass through reactive silanes, but where no reaction is possible a definite maximum in laminate properties is obtained with silanes that have optimum compatibility with the polymer, as predicted from solubility parameters of the organofunctional silane and the resin[56] (Figure 5.7). Coupling through solution compatibility is most successful with glassy polymers like polystyrene, but less effective with crystallizing polymers like polyethylene or polypropylene.

The ultimate in compatibility is obtained with a silane-modified polymer as coupling agent for a similar polymer. Thus, a trimethoxysilane modified polystyrene is a good adhesion promoter for polystyrene, and a vinyltrimethoxysilane-grafted polyethylene is a good primer for polyethylene. A single trimethoxysilyl group with a polymer tail is not sufficient for bonding. In comparing silane–styrene copolymers as coupling agents on glass in polystyrene laminates, it was observed that five to ten trimethoxysilyl groups were necessary in a copolymer of molecular weight of 10,000 for optimum performance (Table 5.14).

**5.4.3.2. Bonding Through Interpenetrating Networks.** Certain reactive silanes with amine, methacrylate, or cationic vinylbenzyl functions often perform very well as coupling agents in thermoplastic composites even

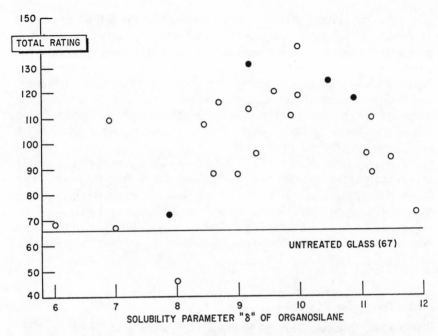

**Figure 5.7.** Relation of solubility parameter of coupling agents with performance of polystyrene–glass composites.

though there is no obvious reaction or preferred solubility of the silane and the polymer.

The performance of methacrylate-, amine-, and cationic styryl-functional silanes as coupling agents for thermoplastic resins becomes clear if we assume formation of interpenetrating polymer networks of siloxanes

**Table 5.14.** Polystyrene Laminates with Polymeric Silane-Treated Glass

| Styrene copolymer coupling agent | | Flexural strength, MPa | |
| Molecular weight | Si atoms/mol | Dry | Wet (2-h boil) |
|---|---|---|---|
| Control | — | 226 | 140 |
| 1,100 | 1 | 250 | 212 |
| 5,400 | 1 | 254 | 170 |
| 10,000 | 1 | 257 | 178 |
| 10,000 | 2 | 250 | 212 |
| 10,000 | 5 | 303 | 238 |
| 10,000 | 11 | 302 | 307 |
| 10,000 | 20 | 308 | 312 |

with thermoplastic resins at the mineral interface. The uncured siloxane film deposited on a mineral surface acts as a lubricant and dispersing aid during initial mixing of fillers with the resin. At molding temperature, the siloxane is a partial solvent for molten thermoplastic, but as the composite cools, the thermoplastic loses its solubility and separates as an interpenetrating phase with the siloxane at the interface. Some cross-linking may occur in the siloxane phase at mold temperatures. Addition of peroxides facilitates this cross-linking of siloxane and may aid in trapping the dissolved thermoplastic phase.

Although interpenetrating polymer networks have not been identified in the interphase region of these silane-modified composites, there are a number of such examples in related systems. Interpenetrating polymer networks have been observed in silane-modified polymer films deposited from a latex.[57] Such modified latex films have improved adhesion to mineral substrates and improved solvent resistance, even though there is no chemical reaction between the siloxane and the organic polymer film.

Cold blends of trackifying resins with small proportions of aminoorganofunctional silanes are effective in improving the adhesion of thermoplastic elastomers to mineral surfaces.[58] The same resins and silane coupling agents used separately were relatively ineffective as adhesion promoters. Chemical reactions between silane and tackifying resin or between the silane–resin mix and the elastomer do not appear to be necessary.

Tackification of thermoplastic elastomers is believed to require partial compatibility of a low-molecular-weight resin to provide domains of resin dispersed in a rubbery matrix. Since the compatibility of a resin in the thermoplastic is modified by adding a silane, an optimum balance of silane and resin must be determined for each elastomer.

The silane-tackifier concept should be applicable with rigid thermoplastics if the proper silane–resin combination can be found that will form interpenetrating polymer networks with the thermoplastic during molding and cooling. The resin must be of low enough molecular weight to provide a mobile liquid under molding conditions. It must have some compatibility with the hot thermoplastic, yet separate as an interpenetrating phase upon cooling. Some cross-linking or grafting to the thermoplastic may be beneficial.

A polymerizable amine-functional silane (H of Table 2.2) activated with dicumyl peroxide is a very effective adhesion promoter for hydrocarbon polymers.[59] A silane-modified hexa(methoxymethyl) melamine resin primer was most effective in bonding polar thermoplastic polymers to glass (Table 5.15). In both of the above cases, it was postulated that the silane-modified primer acted as a solvent for the molten thermoplastic polymer in the interphase region, but cooled to an interpenetrating polymer network.

**Table 5.15.** Adhesion of Thermoplastics Fused against Printed Glass

| Thermoplastic | | Primer on glass | |
|---|---|---|---|
| Type | Name | Silane peroxide[a] | Silane melamine[b] |
| Polyester (elastomeric) | Hytrel 4055[c] | poor | excellent |
| Polyester (rigid) | Valox 310[d] | fair | good |
| Polyurethane (elastomeric) | Estane 58300[e] | fair | excellent |
| Polyamide | Zytel 61[c] | excellent | excellent |
| Polycarbonate | Lexan 141 112[d] | good | good |
| Polyether (polyblend) | Noryl[d] | fair | excellent |
| Polystyrene | Styron 636[f] | excellent | poor |
| Polypropylene | Profax 6323[g] | excellent | poor |
| Polyethylene | | excellent | poor |
| EVA | Elvax 150[c] | excellent | nil |

[a] 1% dicumyl peroxide in silane H (Table 2.2) diluted to 10% solids in methanol.
[b] 10% silane C (Table 2.2) in Cymel 303® (Amer. Cyanamid).
[c] Product of DuPont.
[d] Product of General Electric.
[e] Product of B. F. Goodrich.
[f] Product of Dow Chemical Company.
[g] Product of Hercules.

A silane primer that interpenetrates different polymers and bonds them to glass should interpenetrate such polymers brought into intimate contact and bond them to each other. The primers of Table 5.15 were used to bond a slab of polyethylene to a polyester film (Mylar®) through holt-melt adhesives.[60] Either an EVA terpolymer (Dupont® CXA 1025) or a thermoplastic polyester elastomer (Goodyear® Vitel 5571) were good hot-

**Table 5.16.** Bonding Polyethylene to Mylar® Through Hot Melts at 150°C

| "Hot Melt" Polymer | Primer on PE | Peel Str. (N/cm) PE | Primer on Mylar® | Peel Str. (N/cm) to Mylar® |
|---|---|---|---|---|
| CXA 1025[1] | None | 1.8 | None | 0.8 |
| CXA 1025 | A | * | B | * |
| Vitel 5571[2] | None | 3.5 | None | 11.5 |
| Vitel 5571 | A | * | B | * |

* Cohesive failure in adhesive at greater than 30 N/cm.
[1] EVA terpolymer, product of DuPont.
[2] Elastomeric polyester, product of Goodyear.
Primer A = 1% dicumyl peroxide in silane H (10% in methanol).
Primer B = 10% silane C in Cymel® 303 (10% in isopropanol).

melt adhesives to the primed surfaces. Adhesion to unprimed polyethylene or polyester were relatively poor (Table 5.16).

## References

1. A. J. Bush. *Modern Plast.* **35**, 143 (1958).
2. F. J. McGarry and M. Fyiwara. *Modern Plast.* **45**, 143 (1968).
3. L. J. Broutman. SPI, 25th Ann. Tech. Conf. Reinf. Plast. 13-B (1970).
4. M. L. Hair. *Infrared Spectroscopy in Surface Chemistry.* Marcel Dekker, New York (1967).
5. R. G. Pike and D. Hubbard. *J. Res. Nat. Bur. Stand.* **591**, 127 (1957).
6. J. J. Friapat, A. Jelli, G. Poncelet, and J. Andre. *J. Phys. Chem.* **69**, 2185 (1965).
7. W. Hinz. *Glastech. Ber.* **31**, 422 (1958).
8. G. B. Cvijanovich. *Found. Phys.* **7**(11/12), 785 (1977).
9. K. Gutfreund and H. S. Weber. SPI, 16th Ann. Tech. Conf. Reinf. Plast. 8-C (1961).
10. B. M. Vanderbilt. SPE, 22nd ANTEC 23-1 (1966).
11. T. Yamaguchi, T. Ono, and H. Ito. *Angew. Makromol. Chem.* **32**, 177 (1973).
12. S. Sterman and J. G. Marsden. *Plast. Tlchnol.* **9**, 39 (1963).
13. R. C. Osthoff, A. M. Bueche, and W. T. Grubb. *J. Am. Chem. Soc.* **76**, 4659 (1954).
14. E. H. Andrews, H. Pingsheng, and C. Vlachos. *Proc. R. Soc. London A* **381**, 345 (1982).
15. D. L. Chamberlain, M. V. Christensen, and M. Bertolucci. SPI, 24th Ann. Tech. Conf. Reinf. Plast. 19-C (1969).
16. L. Pauling. *The Nature of the Chemical Bond,* 2nd ed., p. 69, Cornell University Press, Ithaca, N.Y. (1948).
17. W. Hertl. *J. Phys. Chem.* **72**, 1248 (1968).
18. P. Dreyfuss. *Macromolecules* **11**(5), 1031 (1978).
19. P. Dreyfuss, L. J. Fetters, and A. N. Gent. *Macromolecules* **11**(5), 1036 (1978).
20. P. W. Erickson. SPI, 24th Ann. Tech. Conf. Reinf. Plast. 19-B (1969).
21. D. J. Vaughan and E. L. McPherson. SPI, 27th Ann. Tech. Conf. Reinf. Plast. 21-C (1972).
22. E. P. Plueddemann. SPI, 27th Ann. Tech. Conf. Reinf. Plast. 21-B (1972).
23. E. R. Pohl and F. O. Osterholz. In *Silanes, Surfaces, and Interfaces,* D. E. Leyden, Ed., pp. 481–500, Gordon and Breach, Amsterdam, New York (1986).
24. H. A. Clark and E. P. Plueddemann. *Mod. Plast.* **40**(6), 133 (1963).
25. H. Ishida and J. L. Koenig. *J. Polym. Sci. Polym. Phys. Ed.* **17**, 615 (1979).
26. R. L. Kaas and J. L. Kardos. SPE, 32nd ANTEC, Paper 22 (1976).
27. E. P. Plueddemann. *Chemical Bonding Technology for Terrestial Solar Cell Modules,* JPI Report 5101–5132, Sept. (1979).
28. E. P. Plueddemann. In *Molecular Characterization of Composite Interfaces,* H. Ishida and G. Kuman, Eds., p. 17, Plenum, New York (1985).
29. B. N. Vanderbilt. SPI, 17th Ann. Tech. Conf. Reinf. Plast. 10-D (1962).
30. K. H. G. Ashbee and R. C. Wyatt. *Proc. R. Soc. London A.* **312**, 533 (1969).
31. H. Ishida. In *Molecular Characterization of Composite Interfaces,* H. Ishida and G. Kuman, Eds., p. 34, Plenum, New York (1985).
32. E. P. Plueddemann and P. G. Pape. SPI, 40th Ann. Tech. Conf. Reinf. Plast. 17-F (1985).
33. E. P. Plueddemann. *Adhes. Age* June, 36 (1975).
34. E. P. Plueddemann. *Proc. Third Ann. Conf. Adv. Composites,* p. 283, ASM Intl. Sept. (1987).
35. E. P. Plueddemann and P. G. Pape, SPI, 42nd Ann. Tech. Conf. Reinf. Plast. 21-E (1987).
36. Y. Ekstein. *J. Adhes. Sci. Technol.* **3**(5), 317 (1989).

37. L. D. Tryson and J. L. Kardos. SPI, 36th Ann. Tech. Conf. Reinf. Plast. 2-E (1981).
38. J. Jang, H. Ishida, and E. P. Plueddemann. SPI, 44th Ann. Tech. Conf. Reinf. Plast. 9-B (1989).
39. A. C. Zettlemoyer. In *Chemistry and Physics of Interfaces*, D. E. Gushee, Ed., Chapter XII, American Chemical Society, Washington, D.C. (1965).
40. J. C. Bolger. In *Adhesion Aspects of Polymeric Coatings*, K. L. Mittal, Ed., pp. 3–18, Plenum, New York (1983).
41. F. M. Fowkes. *J. Adhes. Sci. Technol.* 1(1), 7 (1987).
42. F. J. Boerio and R. G. Dillingham. In *Adhesive Joints*, K. L. Mittal, Ed., p. 541 Plenum, New York (1984).
43. F. J. Boerio, R. G. Dillingham, and R. C. Bozian. SPI, 39th Ann. Tech. Conf. Reinf. Plast. 4-A (1984).
44. P. Walker. *J. Coatings Technol.* 52(668), 33 (1980).
45. E. P. Plueddemann. *Interfaces in Polymer, Ceramic, and Metal Matrix Composites*, H. Ishida, Ed., pp. 17–34 Elsevier, New York (1988).
46. G. A. Parks. *Chem. Rev.* 65, 127–198 (1964).
47. L. Pauling. *The Nature of the Chemical Bond*, Cornell University Press, Ithaca, N.Y. (1939).
48. J. Kragton. *Atlas of Metal–Ligand Equilibria in Aqueous Solution*, Halstead Press, New York (1978).
49. M. K. Chandbury, T. M. Gentle, and E. P. Plueddemann. *J. Adhes. Sci. Technol.* 1(1), 29–38 (1987).
50. D. R. Coulter, E. F. Cuddihy, and E. P. Plueddemann. Chemical Bonding Technol. for Terres. Photovoltaic Modules, 5101-232, DOE/JPL-1012-91 (Feb. 1983).
51. E. P. Plueddemann and G. L. Stark. SPI, 28th Ann. Tech. Conf. Reinf. Plast. 21-E (1973).
52. E. P. Plueddemann. SME Conference, Dearborn, Mich., Nov. (1989).
53. E. P. Plueddemann. *J. Adhesion Sci. Technol.* 3(2), 131 (1989).
54. L. H. Sperling. In *Recent Advances in Polymer Blends, Grafts, and Blocks*, L. H. Sperling, Ed., p. 93, Plenum Press, New York (1974).
55. Y. K. Lee and J. D. Craig. The Electrochemical Society 159th Meeting Minneapolis, MN, Paper 141 (1981).
56. E. P. Plueddemann. *Modern Plast.*, 54, 102 (1966).
57. M. A. Lutz and K. E. Palmanteer. *J. Coatings Technol.*, 51(652), 37 (1979).
58. E. P. Plueddemann. *Adhesives Age*, 18(36), June (1975).
59. E. P. Plueddemann and G. L. Stark. SPI, 35th Ann. Tech. Conf. Reinf. Plast. 13-A (1980).
60. E. P. Plueddemann. In *Surface and Colloid Science in Computer Technology*, K. L. Mittal, Ed., p. 152, Plenum Press, New York (1987).

# 6 | Performance of Silane Coupling Agents

## 6.1. General

The need for coupling agents was recognized in 1940 when glass fibers were first used as reinforcements in organic resins. Specific strength-to-weight ratios of dry glass–resin composites were very favorable, but the laminates lost much of their strength during prolonged exposure to moisture. Because unsaturated polyester resins were the most common organic matrix material, various unsaturated compounds of silicon and other elements were tested as coupling agents. Only unsaturated silanes and methacrylato-chrome complexes (DuPont Volan®) have reached commercial importance.

As fiberglass reinforcements were used with other resins, new coupling agents were introduced with specific reactivity for these resins. In general, the best silane coupling agents are those where the organofunctional group on silicon has maximum reactivity with the particular thermosetting resin during cure. Silane coupling agents were also developed for reinforced elastomers and thermoplastic resins.

The same silane coupling agents that were used on glass were also effective on particulate minerals in filled polymers (Chapter 7), and as primers on metal or ceramic surfaces for adhesion of sealants, paints, and adhesives. In some cases, silane coupling agents even promoted adhesion of an organic polymer to another organic polymer. Representative commercial silanes of Table 6.1 are discussed below in their applications with various types of polymers (Figure 6.1).

## 6.2. Unsaturated Polyesters

Fiberglass manufacturers apply a silane-containing size to fiberglass immediately as the fibers are pulled from the melt. Compositions of these

**Table 6.1.** Representative Commercial Coupling Agents

| Organofunctional group | Chemical structure |
|---|---|
| A. Vinyl | $CH_2{=}CHSi(OCH_3)_3$ |
| B. Chloropropyl | $ClCH_2CH_2CH_2Si(OCH_3)_3$ |
| C. Epoxy | $\overset{\displaystyle O}{\overset{\displaystyle /\backslash}{CH_2CHCH_2OCH_2CH_2CH_2Si(OCH_3)_3}}$ |
| D. Methacrylate | $\overset{\displaystyle CH_3}{\overset{\displaystyle \mid}{CH_2{=}C{-}COOCH_2CH_2CH_2Si(OCH_3)_3}}$ |
| E. Primary amine | $H_2NCH_2CH_2CH_2Si(OC_2H_5)_3$ |
| F. Diamine | $H_2NCH_2CH_2NHCH_2CH_2CH_2Si(OCH_3)_3$ |
| G. Mercapto | $HSCH_2CH_2CH_2Si(OCH_3)_3$ |
| H. Cationic styryl | $CH_2{=}CHC_6H_4CH_2NHCH_2CH_2NH(CH_2)_3Si(OCH_3)_3 \cdot HCl$ |
| I. Cationic methacrylate | $\overset{\displaystyle CH_3}{\overset{\displaystyle \mid}{CH_2{=}C{-}COOCH_2CH_2{-}\overset{\displaystyle Cl^- }{\underset{}{N(Me_2)}}^{\oplus}CH_2CH_2CH_2Si(OCH_3)_3}}$ |
| J. Chrome complex | |
| K. Titanate | $\overset{\displaystyle CH_3}{\overset{\displaystyle \mid}{(CH_2{=}C{-}COO)_3TiOCH(CH_3)_2}}$ |
| L. Cross-linker | $(CH_3O)_3SiCH_2CH_2Si(OCH_3)_3$ |
| M. Mixed silanes | $C_6H_5Si(OCH_3)_3 + F$ |
| N. Formulated | Melamine resin + C |

sizes are proprietary and are merely described as "compatible" with specific resins. Adhesion promotion by silanes is modified with other materials to provide fiber protection, lubrication, antistatic properties, strand integrity, and wet-out by resins.

The performance of coupling agents may be tested by treating heat-cleaned glass cloth with aqueous dispersions of individual coupling agents. Typical polyester laminates prepared with 12 plies of glass cloth treated with 0.5% aqueous coupling agent are described in Table 6.2. The methacrylato-chrome complete was a favorite for years, and the typical green color imparted to laminates was associated with high-quality polyester laminates.

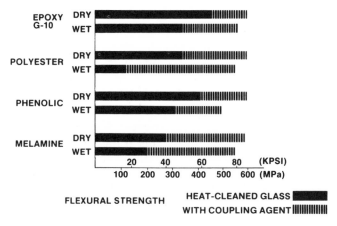

**Figure 6.1.** Effect of coupling agents on flexural strengths of glass cloth-reinforced thermosetting resins.

The methacrylate ester silane is preferred with most general purpose polyesters. The cationic vinylbenzyl silane gives best performance with vinyl resins (acrylic-modified epoxies). It also was the one silane uniquely effective in bonding UV-cured vinyl resins to quartz fibers used in fiber optics.[1] It is also recommended in anaerobic curing resins.

Over 100 unsaturated organosilicon compounds were compared as coupling agents on glass in polyester laminates.[2] A number of these materials were very effective as coupling agents, but showed no advantage over the silanes of Table 6.2.

**Table 6.2.** Fiberglass-Reinforced Polyester Composites

| | | Flexural strength of composite (MPa) | |
|---|---|---|---|
| Couplimg agent on glass | | | |
| Function | Ref. Table 6.1 | Dry | After 2-hr water boil |
| None | | 386 | 234 |
| BJY finish[a] | | 441 | 386 |
| Chrome complex | J | 503 | 428 |
| Vinyl silane | A | 462 | 414 |
| Methacrylate silane | D | 620 | 586 |
| Cationic methacrylate silane | I | 620 | 566 |
| Cationic vinylbenzyl silane | H | 634 | 566 |

[a] Equimolar $CH_2{=}CHSiCl_3$ and $CH_2{=}CCl{-}CH_2OH$.

Rather than look for new organofunctional silanes for coupling with polyester resins, efforts in recent years have been aimed toward more effective application of presently available materials.

The effect of pH of aqueous silane baths was shown to be most important with the cationic silanes. Optimum pH for application of coupling agent to glass or silica was at the isoelectric point of the mineral surface (see Figure 4.6).

Neutral organofunctional silanes benefit from addition of an amine catalyst to aid in condensation of silanols at the interface (Table 5.3). An amino-organofunctional silane may be used in admixture with the methacrylate ester-functional silane. The amine acts as catalyst for bonding of silanols to glass and then deposits as a mixed siloxane on the surface. Single silanes were compared with mixed silanes as treatment on glass microbeads in polyester catings (Table 6.3). The diamine-functional silane (F of Table 2.2) was effective as a catalyst as shown by initial properties of the castings, but it caused a decrease in strength retention after a 24-hr water boil. Unreacted amine functions apparently are hydrophilic enough to favor entrance of water to the interphase region. The vinylbenzyl cationic silane H also showed catalytic activity in improving initial bonding of coupling agents to glass, but then participated in polymer cure to provide significant improvement in resistance to a 24-hr water boil.

Flexible polymers with polymerizable plasticizers may be treated like unsaturated polyesters. Reactive unsaturated silanes may be used as primers or as additives to provide water-resistant adhesion to mineral surfaces. Silanes D and H are recommended in PVC plastisols having polymerizable plasticizers. Clear ethylene-vinylacetate (EVA) or ethylene-methacrylate (EMA) copolymers may be cross-linked with small proportions of peroxides to obtain clear, noncreeping encapsulants for solar cells. A mixture of an

**Table 6.3.** Glass Microbead[a]-Filled Polyester Castings

| Coupling agent on glass | | Flexural strength of castings, MPa | | |
|---|---|---|---|---|
| Nature | Ref. Table 6.1 | Dry | 2-hr boil | 24-hr boil |
| Untreated | — | 72 | 54 | 33 |
| Vinyl silane | A | 82 | 76 | 41 |
| Methacrylate silane | D | 145 | 123 | 67 |
| Vinylbenzyl silane | H | 154 | 118 | 62 |
| 9/1 mix | D/F | 150 | 131 | 55 |
| 9/1 mix | D/H | 155 | 135 | 82 |

[a] Potters® 3000 microbeads, 30–50μ.

**Table 6.4.** Adhesion of Cross-Linkable EVA to Various Surfaces
(cured for 20 min at 150°C)

| Primer system | Peel strength after 4 days in water, N/cm | | | | |
|---|---|---|---|---|---|
| | Glass | Copper | Aluminum | Phenolic | Tedlar© |
| Unprimed control | Nil | Nil | 1.0 | 3.9 | Nil |
| Prehydrolyzed D | a | 10.0 | 1.0 | 3.9 | a |
| 1% BDMA in D | a | a | a | a | a |
| *Primer milled into EVA* | | | | | |
| Prehydrolyzed D | 15.6 | 3.9 | 7.8 | 11.7 | 3.9 |
| 9/1 D/E | 19.6 | a | a | a | a |
| 1% BDMA in D | a | a | a | a | a |

[a] Cohesive failure in EVA at about 35 N/cm.

amine with silane D was very effective as a primer or as an additive for bonding to various surfaces[3] with cross-linkable EVA (Table 6.4).

Thin films of unsaturated polyester resin, cured on primed glass or metal coupons (cf. Section 1.3) were used to screen coupling agents of Table 6.1 for relative effectiveness in promoting water-resistant bonds (Table 6.5). Primers were rated according to the improvement in time to failure compared to unprimed controls:

$-$ = *not recommended*, no significant improvement

$+$ = *fair to good*, 10× to 100× improvement

$++$ = *very good*, 100-fold or more improvement.

**Table 6.5.** Coupling Agents for Unsaturated Polyesters (screening tests)[a]
(catalyzed CoResyn® 5500 cured 30 min at 130°C)

| Primer on surface | Retention of adhesion on | | | |
|---|---|---|---|---|
| | Glass | Aluminum | C.R. steel | Titanium |
| J | + | + | + | − |
| K | + | + | − | + |
| APG-2[b] | + | − | − | − |
| D | ++ | ++ | + | + |
| H | ++ | ++ | ++ | ++ |

[a] (−) not recommended; (+) fair–good; (++) very good.
[b] APG-2 = Caveco-Mod® alumino-zirconate, Cavedon Chemical Co.

## 6.3. UV-Curable Coatings

Ultraviolet-light-curable coatings are generally solvent-free liquid oligomers with multiple acrylic functionalities along with a UV absorber to act as initiator. Methacrylates are not initiated by UV in such systems, but methacrylates will copolymerize with activated acrylic monomers. Silane D (Table 6.1) is generally useful as an adhesion-promoting additive or primer for UV-cured acrylic oligomers. A commercial acrylic-modified urethane, Lightweld® 415 (supplied by Dyman Co., Torrington, CT) was tested as thin films cured on various primed surfaces. Silane D proved to be an effective adhesion promoter, but it was improved greatly by mixing with crosslinker L (Table 6.6).

UV-light-cured cyclo-aliphatic epoxide formulations are available under the trade names Cyracures® from Union Carbide Corp. (Danbury, CT). These materials use arylsulfonium salts as initiators which are stable under normal conditions but release cationic initiators ($BF_3$ or $PF_5$) when exposed to UV light.[4] In the presence of the generated cationic species, very rapid polymerization takes place. Thin films of a fomulation comprising 9 parts Cyracure® Resin UVR-610 and 4 parts Cyracure® Initiator UV16990 on glass, aluminum, and steel were passed under twin UV lamps (intensity 30.92 $MW/cm^2$) through a 61.25-cm path at a rate of 3 meters/min. Films were tackfree but were generally given an after cure of 1 hr at 100°C. Of the standard silanes tested, all aminofunctional silanes were completely unsuited since they inhibited the cure. Of the other silanes, several showed significant improvement in adhesion to glass. Unprimed adhesion to aluminum was fairly good and further improved by adding 1% silane to the formulation. None of the films retain adhesion very well to glass or steel in 70°C water (Table 6.7). Fluoride ions derived from the initiator may

**Table 6.6.** Adhesion of UV-Curable Acrylic Coating to Primed Surfaces

| Primer on surfaces | Glass | | C.R. Steel | | Stainless steel | | Aluminum | | Polyester film |
|---|---|---|---|---|---|---|---|---|---|
| | Dry | Wet[a] | Dry | Wet | Dry | Wet | Dry | Wet | Dry (90° peel) |
| None | Good | Nil | Fair | Nil | Good | Nil | Fair | Nil | Poor |
| Silane D | Exc.[b] | Fair | Exc. | Poor | Exc. | Exc. | Exc. | Exc. | Fair (7.5 N/cm) |
| 1/1 Mix D/L | Exc. | Exc. | Exc. | Exc. | Exc. | Exc. | Exc. | Exc. | Fair (7.5 N/cm) |

[a] wet = after 1 day in 70°C water.
[b] exc. = could not be peeled, 100% cohesive failure.

be detrimental for bonding to glass and steel while they may benefit adhesion to aluminum by producing an insoluble aluminum fluoride surface.

## 6.4. Epoxy Resins

### 6.4.1. General

Epoxy resins are a broad class of resins that cure through some reaction of oxirane functionality. The basic resin may be a family of bis-phenol-A glycidyl ether oligomers, novalac glycidyl ethers, aliphatic glycidyl ethers, hydantoin glycidyl ethers, cycloaliphatic epoxides, and others, Brominated modifications of these may be used in fire-retardant composites. Various curing agents used with these resins include polyfunctional aliphatic amines, aliphatic amine polyamides, aromatic amines, cyclic polyacid anhydrides, dicyandiamide, $BF_3$-amine complexes, and others.

A large number of organofunctional silanes are effective with epoxy resins,[2] but certain generalizations can be made in selecting the optimum silane for a given system. The coupling agent should have a reactivity at least comparable to that of the epoxy resin with the particular curing system used. The glycidoxypropyl silane (C of Table 6.1) is an obvious choice in any given glycidylfunctional epoxy resin.

Room-temperature-curing epoxies get maximum performance by addition of primary aminofunctional silanes (E and F). These same silanes should be avoided with anhydride-cured epoxies. A large proportion of the primary amine function is consumed in forming cyclic imides with the curing agent.[5] The mechanism of coupling agent reaction is not always

**Table 6.7.** Adhesion of UV-Curable Epoxy on Unprimed Surfaces, UV + 1 hr (100°C post-cure)

| 1% additive to formulation | Time for loss of adhesion in 70°C water | | |
|---|---|---|---|
| | Glass | Aluminum | C.R. steel |
| None | 15 min | 6 hr | 30 min |
| Silane D | 30 min | $a$ | 1 hr |
| Silane C | 2 hr | $a$ | 1 hr |
| Silane B | 2 hr | $a$ | 1 hr |
| Silane A | 2 hr | $a$ | 2 hr |
| Silane L | 6 hr | $a$ | 1 hr |

$a$ No failure in 4 days in 70°C water.

obvious. Silane B with a chloropropyl functionality is a creditable coupling agent with high-temperature-cured epoxies.[6] Under these conditions, the chloropropyl group can react with aromatic amines or with carboxyl groups (in the presence of benzyldimethyl-amine catalyst).[7]

### 6.4.2. Thin Film Tests

Coupling agents were screened as primers for typical epoxy formulations by comparing the time to failure of thin films of polymer cured on primed microscope slides and soaked in water until interfacial failure could be induced by prying with a razor blade.

Fisher brand 12-550-11 precleaned microscope slides were used directly from the box. Metal coupons were scrubbed with Scotchbrite[TM] pads or with kitchen cleanser on a moistened cloth. Metal surfaces were considered to be clean if water spread evenly.

Primers were prepared by diluting commercial silanes to about 20% active material in isopropanol. Neutral silanes were catalyzed by adding 1% diaminofunctional silane (F) based on total silane. A 2% solution of aluminum secondary butoxide[9] (Chattem Chemical) in toluene was compared with silane primers for some epoxides. Kimwipes[TM] tissues paper (Kimberly Clark) was moistened with primers and wiped on the microscope slides or metal coupons to be tested. After 15 min at room temperature, an epoxy formulation was troweled on the primed surface with a spatula to give films about 0.2–0.5 mm thick. After curing the epoxy, as indicated, the composite was placed in water (70°C or boiling) and tested periodically by attempting to loosen the film with a razor blade. Primers were rated according to the length of time the film retained adhesion. In parallel tests, an epoxy film on silane-primed aluminum showed first signs of failure after 4 hr in 70°C water. The same system in a lap shear test showed only a trace of interfacial failure after 150 days in 70°C water at a lap shear strength of 5000 psi (34.5 MPa).

Ratings of silane primers were:

-    *Not recommended*, although performance may have been better than a control;
- +  *Good*, significantly better than the control, as much as 100-fold increase in time of failure;
- ++  *Very good*, up to 1000-fold improvement over the control; and
- +++  Outstanding, best of the series.

**Table 6.8.** Silane Primers for Epoxies on Glass

| Primer | DETA Cure: RT | TEPA 80°C | DMP-30 RT | Versamid RT | Nadic-Me Anh. 150°C | m-PDA 150°C | DADPS 180°C |
|---|---|---|---|---|---|---|---|
| F | − | − | − | + | − | + | − |
| D | ++ | − | + | − | + | ++ | + |
| H | + | ++ | +++ | + | + | ++ | ++ |
| C | +++ | ++ | + | ++ | + | ++ | + |
| G | ++ | + | − | + | − | + | + |
| B | ++ | +++ | + | + | + | +++ | +++ |
| E | − | − | − | − | − | − | − |
| M | ++ | + | ++ | − | + | ++ | + |
| N | ++ | ++ | ++ | +++ | +++ | − | − |
| Al(s-BuO)$_3$ | + | | | | | | − |
| Limits of H$_2$O test | ←————— 7 days at 70°C —————→ | | | | ←—— 14 days at 70°C   7 days boil | | |

Ratings: −, not recommended; +, good; ++, very good; +++, best of series.

When several primers showed no failure to the limits of the test they were all rated ++.

A technical grade of diglycidyl ether of bisphenol A, DER 331 (Dow Chemical Co.), epoxy equivalent weight 177–188, was used in all homemade formulations. Curing agents included DEH-24 (Dow Chemical Co.) triethyl-enetetramine (TETA) used at 13 parts per 100 parts of resin (phr), DEH-26 (Dow Chemical Co.) tetraethylene pentamine (TEPA) used at 14 phr, Versamid-125 (General Mills) amine polyamide used at 100 phr, DMP-30 (Rohm and Haas) trisdimethylaminophenol used at 10 phr, $m$-phenylenediamine ($m$PDA) used at 14 phr, diaminodiphenylsulfone (DADPS) used at 33 phr, and Nadic methylanhydride (Applied Chemical) used at 100 phr.

A few typical epoxy formulations with standard curing agents were tested on glass microscope slides as indicated in Table 6.8. Simple aminofunctional silanes, which are commonly used as adhesion promoters, are not recommended with most of these formulations. A modified amine functional silane (H) or a mixture of $PhSi(OMe)_3$, with an aminofunctional silane (M) is greatly superior for most of the epoxies. Surprisingly good results were obtained with chloropropyl silane (B) and a silane-modified melamine resin (N).

Some of the silanes were compared as primers and as additives to a commercial room-temperature curing epoxy formulation, Magnabond 6388-3 (Magnolia Plastics, Chandler, GA) on glass (Table 6.9). Aminosilanes F and H were much more effective as additives than as primers. The mixed silane M was very effective as a primer but not as 1% additive. The major component [$PhSi(OMe)_3$] in M is not an adhesion promoter unless it is cohydrolyzed with a reactive silane to form mixed siloxane oligomers with reactive sites for bonding.

It has been disclosed that performance of any silane adhesion promoter can be improved by adding a cross-linking silane to it.[10] The compound bis-trimethoxysilylethane, L, was tested with silanes F and C in primers for

**Table 6.9.** Comparison of Silane
Primers to Additives
(Magnabond 6388-3 to glass)

| Silane | Primer | 1% additive |
|--------|--------|-------------|
| F      | +      | +++         |
| H      | +      | ++          |
| G      | +      | −           |
| M      | +++    | −           |
| N      | +      | +           |

**Table 6.10.** Room-Temperature-Cured Epoxy Films on Glass
(hours in 70°C water to adhesion loss)

| Primer on glass | Polyamide | DMP-30 | DEH-24 |
|:---:|:---:|:---:|:---:|
| None | <1 | <1 | <1 |
| F | 1 | 1 | 1 |
| 9/1 F/L | 2 | 3 | 1 |
| 8/2 F/L | 2 | 3 | 1 |
| 6/4 F/L | 3 | 3 | 2 |
| 1/1 F/L | 4 | 4 | 8 |
| 1/9 F/L | a | a | a |
| L | a | 6 | a |
| C | a | 8 | a |
| 9/1 C/L | a | 36 | a |
| 8/2 C/L | a | a | a |

[a] No failure after 1 week in 70°C water.

room-temperature curing epoxies (Table 6.10). With this cross-linker, it was possible to formulate primers that provided bonds that resisted 70°C water for over 1 week. Silane C required less cross-linker than did silane F to provide water-resistant bonds to glass. The cross-linker itself is a creditable adhesion promoter in these systems, although it is not effective alone with some polymers.

Mixtures of cross-linker with diamine-functional silane, F, were used as sole curing agents for epoxy adhesives bonded to glass and to steel. Silane F itself provided better adhesion than the simple aliphatic amine $C_8H_{17}NHCH_2CH_2NH_2$ of comparable functionality and molecular weight. Water resistance of bonds was further improved significantly by adding cross-linker to F curing agent (Table 6.11).

**Table 6.11.** Aminosilane Mixtures as Curing Agents
for Adhesion to Steel and to Glass
(stoichiometric mix with DER-667 cured 4 hr at 70°C)

| Composition of curing agent | Adhesion to coupons in 70°C water | |
|:---|:---:|:---:|
| | CR steel | Glass |
| Aliphatic diamine (no Si) | <4 hr | 2 hr |
| F | 2 days | 1 day |
| 9/1 F/L | 3 days | 2 days |
| 1/1 F/L | 5 days | 2 days |
| 1/9 F/L | >5 days | >5 days |

### 6.4.3. Printed Circuit Boards

The printed circuit industry is based largely on high-quality glass-reinforced epoxy (copper-clad) laminates with initial insulation resistance of over $10^7$ megaohms. Processing into circuit boards involves drilling, degreasing and drying, metallizing and copper-plating (of holes), rinses, stripping, etching, immersion in molten tin, and reflow followed by final cleaning and testing. For testing, the printed boards are subjected to 10-day humidity cycling per method 106 of MIL-STD-202. Twice daily, the temperature is raised to 65°C at a humidity of 90–98%. After completion of the conditioning, each specimen is removed separately and insulation resistance is measured within 1 min or removal. Properly prepared double-sided boards are several orders of magnitude better than the required 500-megaohm insulation resistance (MIL-P-55110) on adjoining surfaces at specified distances.

During recent years, the density of holes in commercial boards has increased dramatically. Spacing between holes has decreased to the point where 2.5 mm between centers of holes is common. With 1-mm-diam holes, the distance between hole walls is only 1.5 mm (Figure 6.2). Failure occurs when acids or other conductive materials migrate along glass fiber bundles to form a conductive path between adjacent holes. The probability of early failure increases as the density of holes on the board increases.

Vaughan and McPherson[11] reported on the effect of adverse conditions on the resin–glass interface in G-10 epoxy composites. Laminates prepared with various coupling agents on glass were tested for mechanical properties after boiling in water for up to 200 hr. Retention of flexural strength (Figure

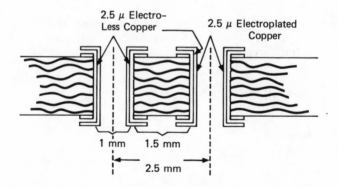

**Figure 6.2.**  Double-sided boards with plated-through holes.

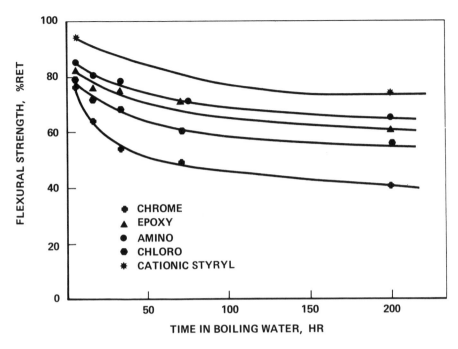

**Figure 6.3.** Histograms showing the retention of flexural strength of epoxy–glass laminates after immersion in boiling water.

6.3) was better with silane coupling agents than with the chrome complex (J). Among the silanes, the cationic styrylfunctional silane (H) was outstanding.

It is somewhat unexpected that a styrene-functional silane and a methacrylate-functional chrome complex are effective coupling agents with dicyanamide cured epoxies (G-10). Reactive double bonds must participate in the epoxy cure reactions. In fact, epoxies are fairly good initiator for double-bond polymerization.[12] The vinylbenzyl cationic silane (H) is among the best coupling agents in G-10 epoxies. A comparable benzyl cationic silane was also a good coupling agent, but not as good as the vinylbenzyl silane H. The vinyl group of H seems to offer additional advantage in the coupling reaction (Table 6.12).

In addition to standard laminate evaluations as shown in Table 6.12 and Figure 6.3, a "pressure cooker" test is valuable in predicting whether completed circuit boards will withstand humidity cycling. Penetration of water into the interfacial region can be accelerated by boiling the coupons in a pressure cooker, and observed by immersing the boiled coupons into

**Table 6.12.** Properties of G-10 Epoxy Laminates
(18 plies of style 7628 fabric)

| Property | Property of laminate | | |
|---|---|---|---|
| Silane on glass | Silane H | Benzyl silane[a] | Epoxy silane C |
| Resin content, % | 37.1 | 37.7 | 39.6 |
| Water absorption @ 23°, % | 0.062 | 0.068 | 0.039 |
| Barcol hardness | 65 | 62 | 60 |
| Flexural strength, MPa | | | |
| A (dry) | 577 | 541 | 520 |
| D (2-hr water boil) | 505 | 469 | 453 |
| Compressive strength, MPA | | | |
| A | 341 | 372 | 377 |
| D | 332 | 332 | 338 |
| Tensile strength, MPa | | | |
| A | 430 | 374 | 363 |
| D | 348 | 325 | 334 |
| Dielectric constant @ 1 Mc/s | | | |
| A | 4.80 | 4.85 | 4.74 |
| D | 4.88 | 4.92 | 4.83 |
| Dissipation factor @ 1 Mc/s | | | |
| A | 0.018 | 0.018 | 0.019 |
| D | 0.019 | 0.019 | 0.021 |

$$\overset{\displaystyle \text{HCl}}{\underset{\displaystyle |}{}}$$

[a] $C_6H_5CH_2NHCh_2CH_2NH(CH_2)_3Si(OCH_3)_3$.

molten solder. Sudden conversion of water in the interior of the laminate to steam causes delamination, which may be observed visually. There is no standard "pressure cooker test," although many laboratories use some modification of this technique to separate "good" laminated from "bad" laminates.

## 6.5. Phenolic Resins

Silane coupling agents may be used to improve the properties of virtually all mineral composites with phenolic resins. Aminofunctional silanes are used with phenolic resin binders on fiberglass insolution, on fiberglass tire cord with resorcinol–formaldehyde resin in RFL dip, and with furan or phenolic resins as binders for sand cores used in metal castings. Aminosilanes have also been proposed with phenolic resins in sand consolidation of oil wells.[13]

Silane coupling agents are commonly used to treat alumina and silicon carbide granules to increase resin wet-out and the mechanical strength of bonded abrasive wheels. Mineral-filled phenolic molding compounds are other obvious areas for improvement through silanes.

The use of silanes as integral blends in phenolic sand core binders illustrates a problem found in a number of silane–resin systems. The silane additives may be reactive with the resin at room temperature, but after storage for even a few hours the silanes lose coupling activity. To be effective, the silanes must be monomeric so they can migrate quickly to filler or reinforcement before the system cures. Premature reaction with the resin reduces mobility such that trace amounts of silane additive becomes ineffective in adhesion promotion. Furane resins, urethanes, and sometimes epoxies resemble phenolic resins in this behaviour. Pretreatment of the filler gives complete utilization of silane adhesion promoters, but is much more costly than simple addition of silane as an integral blend.

"Dog biscuit" test pieces of sand cores were made by mixing 2 parts of catalyzed resin with 98 parts of sand and loading them into steel molds preheated to 232°C. Materials tested were a phenolic, and a furan resin supplied by Core-Lube®. Initial tensile strengths of the test pieces were all in the range of 25–30 MPa, but after exposure to 100% relative humidity at 38°C for 24 hr, strengths dropped to 2–3 MPa, unless an effective adhesion promoter was present. Performance of the diamine-functional silane (F) as adhesion promoter is shown by the wet strength of test pieces with 0.5% silane added to the resin system (Table 6.13). By adding the silane to the catalyst, stable two-part resin systems were obtained. The silane could also be added as a third component at the time of mixing, but this would be very inconvenient in field applications.

**Table 6.13.** Wet[a] Tensile Strengths (MPa) of Sand Core Test Pieces

| 0.5% Silane (F) additive | Wet tensile strength, MPa | |
| --- | --- | --- |
| Conditions | Furan resin | Phenolic resin |
| No silane, control | 2.7 | 2.4 |
| In resin, immediate | 17.8 | 8.6 |
| In resin, 2 hr old | 13.5 | 8.8 |
| In resin, 9 days old | 9.7 | 3.2 |
| In catalyst, 46 days old | 17.9 | 9.2 |
| On sand (none in resin) | 21.1 | 9.4 |

[a] After 24 hr at 38°C and 100% relative humidity.

## 6.6.  High-Temperature Resins

Cross-linked thermosetting resins are infusible after curing and tend to retain useful mechanical properties up to temperatures approaching their thermal breakdown temperatures.

Aliphatic silane coupling agents have heat stabilities comparable to those of conventional thermosetting polyesters, epoxies, and phenolics, but lack oxidation resistance at temperatures wihin the operation range of polymides and other high-temperature resins.[14] Flexural-strength measurements on an early fiberglass-reinforced polyimide composite illustrate the loss in coupling performance at high temperatures. Fiberglass cloth was treated with an aminofunctional silane (E) and laminated with Skybond® 709 resin (Monsanto Chemical Company). Dry and wet (2-hr water boil) flexural strengths were measured on the finished laminate and after heat-aging for 100 hr at 250°C and 318°C (Table 6.14). There was little deterioration of the resin in 100 hr at 318°C as indicated by good dry-strength retention, but the silane at the interface must have decomposed, as indicated by loss in wet strength. Exposure of similar laminates for 100 hr at 318°C in an atmosphere of nitrogen caused little deterioration of either dry or wet strength.

Aromatic organofunctional silanes[14] have a useful temperature range at least 100° higher than that of aliphatic silanes, but none is available commercially. Presence of an aromatic group in the silane molecule seems to impart some improved high-temperature resistance, even though the ultimate bond to silicon is through an aliphatic link. Silane H with a vinylbenzyl group was compared with silane E on glass fabric in polyimide (Resin PN-13, Ciba Geigy) laminates. Silane H contributed better initial properties and better retention of properties after 100 hr at 285°C in air[15] (Table 6.15).

Although aminophenyl silane coupling agents show relatively poor heat stability, it has been reported that when this compound is converted

**Table 6.14.**  Flexural Strength Retention of a Polyimide
Laminate (aminosilane (E) on glass)

| Heat treatment | Flexural strength (% of initial) | |
| --- | --- | --- |
| | Dry | 2-hr boil |
| As laminated | 100 | 94 |
| After 100 hr at 260°C in air | 104 | 91 |
| After 100 hr at 318°C in air | 99 | 25 |

**Table 6.15.** Polyimide[a] Laminates with Silane-Treated Style 7628 Glass Cloth (0.5% silane finish on glass)

|  | Silane H | | Silane E | |
| --- | --- | --- | --- | --- |
| Property | Dry | Wet | Dry | Wet |
| Flexural strength, MPa |  |  |  |  |
| Initial | 508 | 492 | 420 | 366 |
| After 100 hr at 285°C | — | 415 | — | 282 |
| Dielectric constant @ 1 Mc/s | 4.56 | 4.80 | 4.80 | 4.95 |
| Dissipation factor @ 1 Mc/s | 0.006 | 0.007 | 0.020 | 0.021 |
| Dielectric breakdown, KV | >70 | >70 | 45 | 40 |

[a] Data of Vaughan.[16]

**Table 6.16.** Thermal Stability of CA-25 and Silane E with Polyimide Coating on Silcon

|  | Peel strengths retained (%) | | | |
| --- | --- | --- | --- | --- |
|  | Air atmosphere | | Nitrogen atmosphere | |
| Heat treatment time at 400°C (hr) | CA-25 | Silane-E | CA-25 | Silane E |
| 0 | 100 | 100 | 100 | 100 |
| 1 | 62 | 38 | 99 | 88 |
| 2 | 52 | 28 | 95 | — |
| 2.5 | — | — | — | 81 |
| 3 | 50 | — | 97 | — |
| 4 | 41 | — | 99 | 75 |

**Table 6.17.** S-Glass Laminates with Kerimid 601® Polimide

|  | Mixed silanes M | | Silane E | |
| --- | --- | --- | --- | --- |
| Properties of laminates | MPa | psi | MPa | psi |
| Flexural strength initial | 544 | 78,900 | 476 | 69,000 |
| 1000 hr at 260°C | 409 | 59,400 | 258 | 37,400 |
| 2000 hr at 260°C | 306 | 44,400 | 134 | 29,400 |
| Flexural modulus, initial | $24 \times 10^3$ | $3.5 \times 10^6$ | $23 \times 10^3$ | $3.4 \times 10^6$ |
| 1000 hr at 260°C | $21 \times 10^3$ | $3.0 \times 10^6$ | $22 \times 10^3$ | $3.2 \times 10^6$ |
| 2000 hr at 260°C | $20 \times 10^3$ | $2.9 \times 10^6$ | $18 \times 10^3$ | $2.5 \times 10^6$ |

**Table 6.18.** Properties of S-Glass Laminates with PEEK Resin
(45 volume percent glass)

| Coupling agent on glass | Flexural strength | | Interlam-shear strength | |
|---|---|---|---|---|
| | MPa | psi | MPa | psi |
| None | 365 | 52.900 | 27 | 3.930 |
| Aminosilane F | 553 | 80.240 | 35 | 5.070 |
| Mixed silanes M | 669 | 97.000 | 37 | 5.360 |

into an oligomeric polyamide that the product designated CA-25 showed good coupling of polyamide films to silicon wafers[16] with greatly improved thermal stability over that obtained with silane E coupling agent (Table 6.16).

Mixed silanes, containing a high proportion of phenyl trimethoxysilane provide siloxane films with thermal stability comparable to phenylsilicone with adequate chemical reactivity to bond to polymers.[17] Preliminary tests in polyamide (Table 6.17) and PEEK (Table 6.18) laminates show mixture M to be far superior to aminofunctional silanes E or F.

## 6.7. Other Polymers

Silane coupling agents were screened as adhesion promoters for thermoplastic polymers by fusing the resins under light pressure to primed microscope slides or metal coupons and rating them according to the retention of adhesion after soaking in water (cf. Section 6.4.2 for rating system).[17] One series of rigid thermoplastics was fused to glass at 200°C to 300°C (Table 6.19). Another series of thermoplastic polymers was fused at 225°C to 400°C to aluminum (Table 6.20). More flexible thermoplastics were used to glass at moderate temperatures and rated similarly (Table 6.21). Most of the thermoplastics may be considered to be nonreactive with silanes at fusion temperatures. The mechanism of adhesion, therefore, must be some form of interdiffusion and cross-linking of the primer to form pseudointerpolymer networks at the interface.

The adhesion of rubber to silane primers on glass microscope slides was correlated with the performance of the same silanes on particulate fillers as reinforcements in the rubbers. Peel strengths of elastomers vulcanized against primed microscope slides are shown in Table 6.22. Silane H was the most universal adhesion promoter primer of commercially available silanes.

**Table 6.19.** Adhesion of Rigid Thermoplastics to Glass

| Primer | Rigid PVC 200°C | HDPP Herc.-6323 250°C | Polystyrene Styron-430 250°C | ABS Dow 275°C | Nylon Zytel 275°C | PPS Ryton 300°C |
|---|---|---|---|---|---|---|
| F | + | − | − | + | + | + |
| D | − | − | − | − | + | − |
| H | + | + + | + + | + + | + + | − |
| C | − | − | − | − | + | − |
| G | − | − | − | − | + | − |
| B | + | + | − | − | + | − |
| E | − | − | − | − | + | − |
| A | − | − | − | − | − | − |
| M | + | − | − | − | + + | + + |
| N | + | − | − | − | + + | + + + |

**Table 6.20.** Adhesion of Rigid Thermoplastics to Aluminum

| Primer | Polyformal Delrin 225°C | PBT Valox-310 275°C | Polycarbonate Lexan 275°C | PEEK film 400°C | Polysulfone 30-100 300°C |
|---|---|---|---|---|---|
| F | − | − | + | − | + |
| D | − | − | + | − | − |
| H | + + | + | + | + | + |
| C | − | + + | + | − | + |
| G | + | − | − | − | + |
| C | − | − | + | − | − |
| E | − | − | + | − | + |
| A | − | − | − | − | − |
| M | − | − | + + | + | − |
| N | − | + + | + + | + + | + + |

## 6.8. Coatings

The same silanes that are recommended in fiberglass-reinforced epoxy composites are effective primers on metals with comparable polymer coatings. A simple silane primer on metal may give performance comparable to more expensive etchings and chemical modification that provides a porous surface on the metal for mechanical interlocking of the polymer.

Very little has been published on specific applications of silanes in commercial coatings. Primer suppliers and coatings manufacturers consider their formulations proprietary and do not disclose use of silanes in the formulations.

**Table 6.21.** Adhesion of Flexible Thermoplastics to Glass

| Primer | PVC plastisol 180°C | EVA CXA-2022 150°C | SBR block Kraton-1102 200°C | Hi M.W. HDPE 240°C | Mod-PE Plexar-6 240°C | Urethane Estane-3702 175°C | Polyester Hytrel-5525 275°C |
|---|---|---|---|---|---|---|---|
| F | + | − | + | − | + | − | − |
| D | − | + + | + + | + + | − | − | − |
| H | + + | − | + + | + + | + + | + + | + |
| C | − | − | − | − | + + | + + | + |
| G | − | + + | + + | − | + | + + | − |
| B | − | + + | + | + | + | + | + + |
| E | + | − | − | − | − | − | − |
| A | − | + + | − | + | + | − | − |
| M | + + | − | + + + | − | + | + + | − |
| N | + | + + | − | − | − | + + | + |

**Table 6.22.** Adhesion of Vulcanizing Elastomers to Glass
(glass treated with 1% aqueous silane)

| Silane treatment on glass | Peel strength of elastomer to glass (N/cm) | | | | | | |
| | Peroxide | Sulfur | SBR | EPDM Natural | Nitrile | Neoprene | Hypalon |
|---|---|---|---|---|---|---|---|
| None | Nil | Nil | Nil | 0.2 | Nil | Nil | 2.1 |
| F | 2.3 | 6.1 | 2.1 | 7.7 | 14.0 | 1.9 | c |
| D | 5.1 | 8.1 | 0.2 | 0.9 | 0.2 | 0.2 | 3.5 |
| H | c | c | 3.5 | 8.4 | 9.6 | 3.9 | c |
| G | 2.6 | 14.9 | 5.3 | 3.5 | 2.1 | 7.0 | 4.6 |
| B | 0.2 | 2.0 | 0.2 | 1.1 | 0.9 | 0.2 | 2.1 |
| E | 3.7 | 10.0 | 1.9 | 2.5 | 0.2 | 0.2 | 10.5 |
| A | 2.0 | 3.9 | 1.6 | 1.2 | 0.2 | 0.2 | 3.2 |

$c$ = cohesive failure in polymer at >30 N/cm.

Walker[18] compared various silane adhesion promoters as primers and as additives in urethane and epoxy paints. A two-part epoxide-polyamide resin, and a two-part aliphatic isocyanate adduct-cured polyester paint were pigmented with rutile titanium dioxide. Adhesion of the coatings to aluminum and steel was measured after soaking for 1500 hr in water at room temperature, and again after recovering for 48 hr at room temperature and humidity. Wet and recovered adhesion were measured by direct pull-off on mild steel on primed, degreased, or sand-blasted surfaces (Table 6.23). Silane primers on degreased steel were often superior to unprimed adhesion on sand-blasted steel. The diamine-functional silane (F) and the mercaptan-functional silane (G) were the preferred primers. Silane primers were also tested on cadmium, copper, and zinc. Several of the silanes promoted good adhesion through 1000 hr of accelerated weathering exposure to an enclosed carbon arc with water spray. It was rather surprising to observe that the methacrylate silane was fairly effective with both urethane and epoxy paints on several of the metals, although the diamine silane (F) gave most consistent improvement in adhesion when used as a primer.

Silanes were also compared as additives in the same paints on unprimed metals. Data of Table 6.24 (on aluminum) indicate that silanes are especially effective in improving recovery of adhesion after the water soak. These data are consistent with the "reversible hydrolytic bond theory" of adhesion (cf. Section 5.2.3).

Some of the most stringent requirements for adhesion of coatings are those for photoresists on oxidized silicon chips used in electronics. Improved

**Table 6.23.** Wet and Recovered Adhesion Values on Mild Steel

| Paint | Silane/treatment (2% silane of Table 6.1) | Wet adhesion, MPa/% detached | Recovered adhesion, MPa/% detached |
|-------|-------------------------------------------|------------------------------|-------------------------------------|
| Polyurethane | Degreased only | 5.6/100 | 6.8/100 |
| | G/Degreased | 5.4/100 | 12.1/100 |
| | F/Degreased | 7.4/90 | 12.5/90 |
| | Sandblasted only | 11.8/95 | 20.8/60 |
| | G/Sandblasted | 23.6/10 | 29.1/0 |
| | F/Sandblasted | 22.7/30 | 29.1/0 |
| Epoxide | Degreased only | 7.2/100 | 10.9/100 |
| | L/Degreased | 17.3/100 | 21.8/90 |
| | C/Degreased | 14.3/100 | 15.6/100 |
| | G/Degreased | 10.9/30 | 15.4/100 |
| | F/Degreased | 28.1/0 | 29.2/10 |
| | Sandblasted only | 9.2/100 | 20.9/100 |
| | L/Sandblasted | 16.3/30 | 31.7/60 |
| | C/Sandblasted | 8.3/70 | 26.9/100 |
| | G/Sandblasted | 17.2/10 | 30.0/45 |
| | F/Sandblasted | 26.3/50 | 27.8/40 |

adhesion of photoresists to silicon was obtained with standard silane coupling agents to prevent undercutting during wet etching, and lifting of the coating during wet development of resists.[19] Even hexamethyldisilazane, which is usually considered a hydrophobing agent, was effective in preventing undercutting during etching. Each silane adhesion promoter must be

**Table 6.24.** Wet and Recovered Adhesion Values on Aluminum

| Paint | % Silane/treatment (Table 6.1) | Wet adhesion, MPa/% detached | Recovered adhesion, MPa/% detached |
|-------|--------------------------------|------------------------------|-------------------------------------|
| Polyurethane | Degreased only | 1.3/100 | Peeled while drying |
| | 0.4% G/Degreased | 11.7/100 | 13.8/30 |
| | 0.2% F/Degreased | 9.5/100 | 14.5/100 |
| | Sandblasted only | 10.4/70 | 15.2/40 |
| | 0.1% F/Sandblasted | 15.6/70 | 25.2/30 |
| Epoxy | Degreased only | 2.2/100 | 12.9/100 |
| | 0.2% G/Degreased | 26.3/0 | 27.3/20 |
| | 0.2% F/Degreased | 25.3/0 | 26.9/0 |
| | Sandblasted only | 7.4/100 | 22.2/20 |
| | 0.2% G/Sandblasted | 25.1/0 | 26.7/5 |
| | 0.2% F/Sandblasted | 28.2/0 | 28.7/0 |

matched with a proper spinning solvent (zero contact angle), proper preparation of the chip, and match the individual photoresist employed.

## 6.9. Specific Primers

Adhesion of organic polymers to mineral surfaces is often obtained by first depositing an intermediate boundary layer as a primer. Silane-modified primers have been developed to promote adhesion of virtually any organic thermoplastic or thermosetting polymer to metals, glass, and other mineral surfaces, Formulations of commercial primers are proprietary, but certain generalizations can be made as guidelines in preparing silane-modified primers for any application[20]:

1. The primer must form a strong boundary layer, either by self-reaction or by reaction with the topcoating.
2. The final boundary layer must be either tacky (visco-elastic) or rigid with sufficient toughness and strength to carry the mechanical load imposed on the composite.
3. The boundary layer must have polar functional groups (preferably on silicon) for bonding to the mineral.
4. The primer film must have partial compatibility with the matrix polymer. This involves initial solution compatibility with the matrix resin during fabrication and may involve chemical reaction with the resin.
5. The primer film must resist complete solution in the topcoat, that is, it must contribute to a boundary layer at the interface. Resistance to solution is especially important with topcoatings applied from organic solvents or with topcoatings that contain large proportions of liquid monomers. A simple method of observing such solution "lifting" of a primer is to incorporate an intense dye in the primer. The dye should not float to the surface of the topcoat during fabrication of the composite.
6. The primer boundary layer must survive any environment (e.g., oxidation or high-temperature degradation) that the topcoat is expected to withstand. Ultraviolet weather resistance may not be required in the primer if used with pigmented topcoats.

In developing primer formulations, it is helpful to first observe film characteristics of the proposed primer. The cutting action of a razor blade on films deposited on microscope slides is very informative. Some monomeric silanes and aqueous solutions of some silanes give films on glass that dry in air to brittle powders that are relatively useless as primers. Initial films should be soft and develop toughness as they are cured. Sometimes coreaction with a topcoat is depended on for developing of cure. The films should be observed during various baking schedules. Films that remain soft and flexible will not develop a water-resistant bond to the mineral. Very hard films may have good adhesion to the substrate, but are so impervious that topcoats cannot develop a bond to them.

Some typical silane primers are described below with proposed applications.

### 6.9.1.  Partially Prehydrolyzed Silanes in a Solvent[21]

50 parts silane D, E, F, or H
50 parts methanol
 5 parts water

The mixture should be allowed to stand for a few hours at room temperature to equilibrate the oligomeric siloxane structures. The product is then stable indefinitely and may be used directly as a primer or diluted with more alcohol to optimum solids content. Partial prehydrolysis is especially effective with mixed silanes such as M of Table 6.1. Mixtures of silane F with A or $CH_3Si(OMe)_3$ form mixed oligomers that form stable solutions in water. Films of these primers are softer and tougher than those deposited from monomeric silanes or aqueous solutions of the same silanes. These primers are recommended for epoxies, phenolics, polyurethanes, melaminealkyds, polyvinylchloride solutions and plastisols, and numerous thermoplastics (especially with silane H). Primers based on silane H may benefit from addition of 0.5 to 1.0 parts of dicumyl peroxide.

### 6.9.2.  Catalyzed Neutral Silanes

99 parts silane A, B, C, D, G, or L
 1 part benzyldiamethylamine or silane F

The undiluted mixture may be used as integral additive in resin systems, or it may be diluted with about 900 parts of methanol to use as a primer. A prehydrolyzed primer may be prepared by adding 10 parts of water and 1 part of acetic acid to 1000 parts of the alcoholic primer solution. The primer is stable in acid solution, but when it is exposed in a thin film the acetic acid volatilizes with initial solvent loss, leaving an amine-catalyzed silanol for reaction with mineral surfaces.

### 6.9.3. Peroxide-Activated Primers

98 parts silane A, D, or H
2 parts dicumyl peroxide

The primer may be diluted with alcohols and then partially prehydrolyzed by adding water as in the examples of Sections 6.9.1 and 6.9.2.

### 6.9.4. Silane-Modified Epoxy[22]

The diamine-functional silane F is diffunctional in reaction with excess epoxy resins in alcoholic solvent. With loss of solvent from primer films, the third aminofunctional position becomes active, causing the primer to cross-link. This provides a one-component epoxy primer formulation with shelf stability of 6 months or more. Such primer films form very water-resistant bonds to metals and ceramic and engineering plastics, and are good primers for epoxies, phenolics, engineering thermoplastics, and urethanes. A typical formulation for an active primer for polyurethanes is:

8.4 parts liquid diglycidyl ether resin, e.g., DER 330® (Dow Chemical Co.)
1.6 parts silane F
90 parts propylene glycol monoethyl ether

### 6.9.5. Silane-modified Melamine Resins

Cold blends of silanes C or G with commercial melamine (or benzoguanamine) resins are very effective primers for epoxies, urethanes, or engineering thermoplastics.[23] These primers are believed to have initial solution compatibility with matrix resins but cross-link (possibly with some correaction with matrix resin) during fabrication of the composite to form an interpenetrating polymer network at the interface. Different degrees of reactivity and solubility may be formulated into a primer by selecting the proper commercial melamine resin.[24]

Melamine may react with up to 6 mol of formaldehyde to form methylol melamines. The methylol groups may condense with loss of water to form oligomers, or the methylol groups may be stabilized by etherification with low-molecular-weight alcohols. Taking into account these variations, melamine resins may be divided into four groups.

The resins in Group I are characterized by high free hydroxyl and a low degree of methylation. High molecular weight is possible. These resins are supplied in water and, consequently, have not been used in solvent-based primers.

Group II resins have relatively the same amount of free hydroxyl as the resins in Group I have. However, they have a higher degree of methylation and, consequently, lower molecular weight. These melamines are soluble in water or active solvents. Usually they are commercially supplied in an active solvent.

Group III resins have a relatively low amount of free hydroxyl and a high degree of methylation. They are basically monomeric, and they require cosolvent to be reducible in water.

The resins in Group IV are basically low-molecular-weight specialty melamines that have alcohols other than methanol reacted into them. They require strong cosolvent to achieve dispersibility in water.

In general, more stable primers are prepared with hexamethoxy-methyl melamine, (Cymel® 303, American Cyanamide), while more reactive primers result from mixtures of silanes with melamine resins like Resimene® 740 (Monsanto), or Cymel® 325 (American Cyanamide). A typical formulation that is effective as primer, or additive for urethanes, and as primer for engineering thermoplastics is:

90 parts Cymel® 303
10 parts silane C
300 parts methanol

When used as an additive to urethane resin systems, the methanol is omitted. This primer is also effective as an adhesion promoter for fluoro polymers such as Teflon® (DuPont).

### 6.9.6. Silane Tackifiers[25]

Cold blends of commercial tackifying resins and silane F in suitable solvents are effective primers for adhesion of thermoplastic elastomers. The same silane, and sometimes other alkoxysilanes, are effective as adhesion-promoting additives in hot-melt adhesive formulations.[26] Various ratios of silane F to resin may be used, although 2 phr is a good initial concentration for testing:

100 parts tackifying resin (e.g., Piccotex® 75, Hercules)
2 parts silane F
300 parts toluene

Applications of the above primers in various composite structures have been described.[3]

**6.9.6.1. Applications in Polymer to Polymer Adhesion.** Because many of the above primers are believed to function by interdiffusion in polymers with subsequent cross-linking to interpenetrating polymer networks (IPNs), it is not surprising to observe that a primer that will bond two different polymers to glass will bond the polymers to each other.[27]

**6.9.6.2.  Bonding TPR to Organic Surfaces.** Because a given silane-modified tackifier may be a good primer for various dissimilar thermoplastic rubbers to mineral surfaces, it was expected that the same primer might improve adhesion of the rubbers to each other. Ethylene-vinylacetate copolymers (Elvax) were fusion-bonded to styrene-butadiene block copolymers (Kraton) as described in Table 6.25. Under optimum conditions, a silane-tackifier primer (No. 6) gave such a strong bond that failure was entirely cohesive within one of the polymers.

**6.9.6.3.  Bonding Polyethylene to Polyester (Mylar) with a Hot-Melt Adhesive.** Two dissimilar polymers can be bonded through an elastomeric adhesive if the polymer surfaces are primed separately with appropriate primers. Polyethylene sheets were primed with primer 3-H, while the polyester was primed with primer 5. The primed sheets were bonded at 150°C with hot-melt adhesives comprising simple elastomeric thermoplastics. No elastomeric thermoplastic was found that would give good adhesion to both unprimed surfaces. Observed peel strengths of typical elastomeric thermoplastics to both unprimed and primed polyethylene and polyester (Mylar®) are summarized in Table 6.26. Primers 3-H and 5 were effective on polyethylene and Mylar® with all the elastomers, but were of no value when reversed, that is, on Mylar® and polyethylene, respectively.

**6.9.6.4.  Bonding to Silicones.** It is difficult generally to bond organic polymers to vulcanized silicone elastomers. Primer F-1 (based on partially prehydrolyzed silane F), and primer F-6 (2% silane F in Piccotex®-75

**Table 6.25.** Elvax® 40[a] Bonded at 150° to Kraton® 1102[b]

| Primer 6 applied to Kraton® 1102 | Static peel strength, N/cm | |
|---|---|---|
| | Initial | Wet 4 days |
| No primer | 26.6 | 16.1 |
| Resin alone | 48.4 | 18.5 |
| 10% silane F in resin | 77.0[c] | 46.2 |

[a] Elvax® = E/VA copolymer product of DuPont.
[b] Kraton® = SBS block copolymer of Shell.
[c] Cohesive failure in Elvax®.

**Table 6.26.** Bonding Polyethylene to Mylar® through "Hot Melts" at 150°C.

| "Hot melt" polymer | Primer on polyethylene | Peel strength to polyethylene, N/cm | Primer on Mylar® | Peel strength to Mylar®, N/cm |
|---|---|---|---|---|
| CXA 1104[a] | None | 49 | None | 0.7 |
| CXA 1104[a] | 3-H | [c] | 5 | [c] |
| CXA 1025[a] | None | 1.8 | None | 0.8 |
| CXA 1025[a] | 3-H | [c] | 5 | [c] |
| CXA 2022[a] | None | 12.2 | None | Nil |
| CXA 2022[a] | 3-H | [c] | 5 | [c] |
| Vitel 5571[b] | None | 3.5 | None | 11.5 |
| Vitel 5571[b] | 3-H | [c] | 5 | [c] |
| Vitel 5571[b] (reverse prime) | 5 | 5.0 | 3-H | 9.6 |

[a] Product of DuPont.
[b] Product of Goodyear.
[c] Cohesive failure in adhesive at >50 N/cm.

tackifying resin) are fairly good primers on some room-temperature vulcanized (RTV) silicone elastomers, but for most silicones it is necessary to activate the surface by flame or plasma treatment. Contacting the silicone surface with a blue gas flame for about 1 sec is sufficient to activate it. Organic polymers such as urethanes, EVA copolymers, styrene-butadiene block copolymers, and others that bond to silane primers on glass will bond to silane primers on flame-treated silicone as shown in Table 6.27. Flame treatment, or primers alone, did not give good adhesion to silicones in these tests.

**Table 6.27.** Bonding to Cured Silicone Elastomers (N/cm peel).

| Silicone | Primer on silicone | Elvax® 150 at 150°C | | Kraton® 1102 at 210°C | |
|---|---|---|---|---|---|
| | | Untreated | Flame treated | Untreated | Flame treated |
| DC 9596 | None | 0.6 | 0.6 | 0.2 | 1.0 |
| DC 9596 | F-1[a] | 1.5 | 23.1 | 1.5 | [c] |
| DC 9596 | F-6[b] | 0.8 | 10.4 | 0.3 | [c] |
| DC 3140 | None | 0.2 | 0.5 | 0.3 | — |
| DC 3140 | F-1[a] | 0.3 | [c] | 0.2 | 5.0 |
| DC 3140 | F-6[b] | 0.4 | 13.5 | 0.3 | 9.6 |

[a] F-1 = partially prehydrolyzed silane.
[b] F-6 = 2% silane in Piccotex® 75.
[c] Cohesive failure in polymer at >30 N/cm.

# References

1. H. N. Vazerami, H. Schonborn, and T. T. Wang. *J. Radiat. Curing* 4(4), 18 (1977).
2. E. P. Plueddemann, H. A. Clark, L. F. Nelson, and K. R. Hoffman. *Mod. Plast.* 39, 136 (1962).
3. D. R. Coulter, E. F. Cuddihy and E. P. Plueddemann. Flat-Plate Solar Array Project, JPL Report 5101-232 (1983).
4. J. Koleski, Soc. Manuf. Eng. Tech. Paper FG-85-798 (1985).
5. C. H. Chiang and J. L. Koenig. SPI, 35th Ann. Tech. Conf. Reinf. Plast. 23-D (1980).
6. E. Lotz, D. Wood, and R. Barnes. SPI. 26th Ann. Tech. Conf. Reinf. Plast. 14-D (1971).
7. R. L. Merker. U.S. Patent 2,793, 223 (to Dow Corning) (1957).
8. E. P. Plueddemann. *J. Adhesion Sci. Technol.* 2(3), 179–188 (1988).
9. R. A. Pike and F. P. Laman. *Proc. ACS Div. Polym. Mater. Sci. Eng.* 56, 299 (1987).
10. E. P. Plueddemann and P. G. Pape. SPI, 42nd Ann. Tech. Conf. Reinf. Plast. 21-E (1987).
11. D. J. Vaughan and E. L. McPherson. SPI, 27th Ann. Tech. Conf. Reinf. Plast. 21-C (1972).
12. E. Behnke. *Kunstst. Rundsch.* 6, 217 (1959).
13. H. H. Spain. U.S. Patent 3,297,086 (to Esso Research) (1967).
14. E. P. Plueddemann. SPI, 22nd Ann. Tech. Conf. Reinf. Plast. 9-A (1967).
15. E. P. Plueddemann and P. G. Pape. SPI, 40th Ann. Tech. Conf. Reinf. Plast. 17-F (1985).
16. G. C. Tesoro, G. P. Rajendran, C. Park, and D.r. Uhlmann. *J. Adhesion Sci. Tech.* 1(1), 39 (1987).
17. E. P. Plueddemann. Proc. Am. Soc. for Composites, First Techn. Conf., Technomic Publ. Co., pp. 264–279 (1985).
18. P. Walker. *J. Coatings Technol.* 52(668), 33 (1980).
19. J. N. Helbert and H. G. Hughes. In *Adhesion Aspects of Polymeric Coatings*, K. L. Mittal, Ed., pp. 499–508. Plenum, New York (1983).
20. E. P. Plueddemann. *J. Paint Technol.* 42(550), 600 (1970).
21. B. M. Vanderbilt and R. E. Clayton. U.S. Patent 3,350,345 (to Esso Research) (1967).
22. E. P. Plueddemann. SPI, 24th Ann. Tech. Conf. Reinf. Plast. 19-A (1968).
23. E. P. Plueddemann. U.S. Patent 4,231,910 (to Dow Corning) (1980).
24. L. A. Rutter. *Paint Varn. Prod.* March 30 (1973).
25. E. P. Plueddemann. U.S. Patent 3,981,851 (to Dow Corning) (1976).
26. T. P. Flanagan and I. Kaye. U.S. Patent 3,644,245 (to National Starch and Chemical Corp.) (1972).
27. E. P. Plueddemann. In *Surface and Colloid Science in Computer Technology*, K. L. Mittal, Ed., pp. 143–153, Plenum, New York (1987).

# 7 | Particulate-Filled Composites

## 7.1. The Total Picture

Silane coupling agents are generally considered to be adhesion promoters between mineral fillers and organic matrix resins and as such provide improved mechanical strength and chemical resistance to the composite.

Although adhesion is central to any "coupling" mechanism, it is recognized that many factors are involved in the total performance of a composite system. Silane modification of the organic–inorganic interface will also produce changes in other properties of the mixture that may, at times, be more important than the final adhesion across the interface.[1] The interface, or interphase region, between polymer and filler involves a complex interplay of physical and chemical factors related to composite performance as indicated in Figure 7.1.

The central area of adhesion determines the mechanical strength and chemical resistance of a composite, assuming that all other factors are controlled properly. This is the area commonly associated with true coupling agents such as the chrome complexes and organofunctional silanes.

The filler surface interacts with the polymer through catalytic activity, orientation of molecular segments, and other modification of polymer morphology. One important function of silane treatments on fillers is to reduce the inhibitory action of fillers on cure of thermosetting resins and to promote alignment of molecular segments of thermoplastic polymers.

Failure in a composite often is in a boundary layer of filler or resin, rather than at the true interface. A silane at the interface may protect a mineral surface against fracture, or may strengthen the boundary layer of

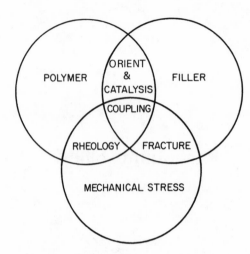

**Figure 7.1.** Interrelationships of polymer, filler, and mechanical stress in composites.

resin to have a positive effect on composite properties even if failure is not at the interface.

Surface modification of fillers also affects the rheology of the polymer mix by changing wet-out, dispersion of particles, viscosity, thixotropy, and flow during plastic fabrication. Surface-active additives have been used with pigments and fillers to control rheology of paints, but very little has been published about such materials in filled plastics.[2]

It is recognized that the total coupling mechanism involves all of these areas and that they are interrelated. Under ideal conditions, a treated filler will wet-out and disperse readily in the plastic with Newtonian flow. The treatment protects the filler against abrasion and cleavage during mixing and in the final composite. The treatment contributes toward strengthening the boundary layer of resin by promoting optimum alignment of polymer segments at the interface and overcoming inhibitory catalytic effects the filler may have on polymer cure. The treated filler should remain chemically inert with the plastic during mixing, but combine with the polymer during the final cure or molding operation.

Commercial organofunctional silanes are available (Table 6.1) with a history of successful application as adhesion promoters in virtually all reinforced composites. Others could be introduced where new types of reactivity are desired, but it would be more profitable to determine methods of using known materials to maximum utility in total control of interface factors—rheology, catalysis, and surface protection—without sacrificing adhesion needed for true "coupling" across the interface.

Optimum performance in all these areas may require some modification of the commercial silanes. This modification may be simple mixing of silanes with film-forming polymers or surface-active agents, or it may be specific chemical modification of the organofunctional group as illustrated in Table 2.3.

## 7.2. Fracture

Improved adhesion between resins and filler may not result in improved mechanical performance of composites if stresses across the interface result in fracture of resin or filler before the interface fails. Several modes of failure are possible.

### 7.2.1. Resin Failure

Organic resins generally have a much higher coefficient of thermal expansion than mineral fillers so that as composites are cooled from elevated fabrication temperatures, shrinkage stresses are set up across the interface. When a thermoplastic resin has a high coefficient of thermal expansion combined with a low elongation at failure, fracture will occur first in the resin phase. Resins that may fall into the category are described in Table 7.1. Simple improvement in adhesion cannot always be expected to impart a significant improvement to composite mechanical properties of these resins with mineral fillers. Some of the value of silane coupling agents may be in modifying polymer properties in the interphase region through an inter-penetrating polymer network to provide greater toughness in the resin. This

**Table 7.1.** Properties of Brittle Resins and Glass Compared

| Polymer | Coefficient of thermal expansion ($c \times 10^6/°C$) | Elongation, % |
|---------|-------------------------------------------------------|---------------|
| Molding acrylics | 50 | 2–10 |
| MMA–styrene copolymer | 60–80 | 3–10 |
| Epoxy | 45–65 | 3–6 |
| Phenolic | 25–60 | 1–1.5 |
| Polyamide–imide | 35–40 | 2.5 |
| Polyphenylene sulfide | 55 | 3 |
| Polystyrene | 60–80 | 1–2.5 |
| Styrene–acrylonitrile | 36–38 | 1.5–3.7 |
| E-glass fibers | 6 | 1.5–3.0 |

explains the observations of DiBenedetto and Scola[3] on adhesive failures
in S-glass polysulfone composites. Aminofunctional silane from the glass
surface was observed up to 100 Å into the polymer matrix. Failure with or
without silane on the glass occurred in the polysulfone within 30 Å of the
glass surface, but shear strength was almost doubled by the presence of
silane coupling agent.

Holtmann[4] observed that good coupling agents on glass protected
polyester matrix resins against hydrolysis when composites were subjected
to prolonged water boil.

Epoxy resins are relatively resistant to hydrolysis, but Vaughan[5]
observed apparent resin damage in some epoxy composites after boiling in
water for 72 hr. A composite made with a poor coupling agent showed voids
in the resin at a fracture surface (Figure 7.2), while a composite made with
a better finish on glass showed no such resin deterioration (Figure 7.3). The
finished glass surface, in the first example, may have inhibited the epoxy-
dicyanidiamide reaction to such an extent that the boundary layer of resin
was undercured. The voids shown in Figure 7.2 may also have resulted
from poor pre-preg preparation.

In a study of fiberglass-reinforced thermoplastic composites[6] a thin
silane-modified rubbery phase was bonded to the mineral surface to provide

**Figure 7.2.**  Glass–epoxy laminate showing voids in the resin after 72-hr water boil.

**Figure 7.3.** Glass–epoxy laminate with silane H coupling after 72-hr water boil.

coupling along with a mechanical cushion to absorb shrinkage stresses. The resulting compression-molded glass composites showed improved strength, toughness, and resistance to aqueous chemicals, but injection moldings failed to show comparable improvements. The high shear involved in injection molding appeared to destroy the interface structure built up through a silane-modified rubber on the glass.

### 7.2.2. Filler Failure

Many of the softer fillers (e.g., clay, talc, alumina trihydrate, and calcium carbonate) may fail along planes of cleavage at stresses below the fracture point of the resin. Improved adhesion will not improve the mechanical properties of composites of such fillers in rigid resins. The value of silane-treated fillers in such systems may be more to improve the rheology in liquid resins or to improve chemical and electrical resistance of composites. Silane adhesion promoters may be very effective, however, in improving mechanical properties of soft filler composites with rubbers or flexible resins where larger proportions of stresses are carried by the resin. As a general rule, it appears that silane treatment of any filler with a Moh hardness of 3 or less will have limited effectiveness in strengthening rigid thermoset or thermoplastic resins, since cleavage will occur in the filler before failure in the resin or the interface (Table 7.2).

**Table 7.2.** Hardness of Fillers Used in Reinforced Plastics

| Soft fillers | | Hard fillers | |
|---|---|---|---|
| Mineral | Hardness, Mohs | Mineral | Hardness, Mohs |
| Talc | 1 | Barite | 4.5 |
| Mica | 2 | Wollastonite | 5 |
| Clay | 2.5 | Glass | 5–6 |
| Calcite | 3 | Quartz | 7 |
| Alumina trihydrate | 3 | Zircon | 7 |
| Dolomite | 3.5 | Corundum | 9 |

Electron micrographs of replicas of fracture surfaces of filled polyesters showed different response of hard and soft fillers to silane coupling agents. Figure 7.4 shows numerous exposed silica particles resulting from interfacial failure in a silica-filled polyester. A casting with silane-treated filler showed fracture (Figure 7.5) entirely in the resin matrix. There was no apparent filler or interfacial failure. After silane pretreatment of a clay filler, a fractured surface showed delamination of clay particles (Figure 7.6). Flexural strengths of silica-filled castings were more than doubled by pre-treating the filler with a methacrylate-functional silane. The same treatment on clay provided only a slight improvement in flexural strength of filled polyester castings.

Glass fibers are especially sensitive to failure from flaws initiated by scratching against one another during fabrication into composites. Silane coupling agents serve to protect glass fibers against scratching, and in some cases a silane finish appears to actually heal flaws on the glass surface.[7] Even though glass is rated as a hard filler, individual glass filaments have been observed to split and rupture when epoxy composites were fractured after boiling in water for 72 hr[4] (Figure 7.7). Other hard fillers such as alumina, silica, zircon, and certain silicates have sufficient strength and respond so well to appropriate coupling agents that fracture of their composites is entirely through the resin with practically no failure at the interface or through filler cleavage. Such fillers, however, are more difficult to grind to fine powders than are the softer fillers, and are more abrasive to molding equipment. The greatest volume of fillers used in plastic composites is therefore made up of softer minerals, even though ultimate strengths obtainable are less than those obtainable with hard fillers. When high strength is required in a composite, soft fillers are often used along with glass fibers for optimum combinations of molding properties and mechanical strength.

**Figure 7.4.** Silica–polyester composite fracture after 2-hr water boil. Large arrows indicate exposed silica particles.

## 7.3. Effects of Filler Surfaces on Polymer Properties

### 7.3.1. Orientation of polymer Segments

Bascom[8] studied the adsorption of organofunctional silane films from water onto stainless steel and glass. The films behaved with respect to contact angle measurements like well-oriented monolayers, even though

**Figure 7.5.** Silica–polyester (with silane H) composite fracture after 2-hr water boil.

ellipsometric measurements left no doubt that the films were polymolecular with thicknesses equivalent to ten or more monolayers. These observations suggest that the first monolayer is covered with well-oriented bilayers.

Molecular orientation may extend beyond the silane layer itself, as shown by orientation of liquid crystals on silane-treated surfaces. In this way, Kahn[9] obtained perpendicular orientation of liquid crystals on surfaces treated with 3-(trimethoxysilyl)propyldimethyloctadecyl ammonium chloride, and parallel alignment on surfaces treated with N-methyl-3-aminopropyltrimethoxysilane.

**Figure 7.6.** Clay–polyester (with silane H) composite fracture after 2-hr water boil.

**Figure 7.7.** Glass–epoxy laminate showing glass rupture after 72-hr water boil.

The effect of silane-modified kaolin filler on orientation of high-density polyethylene in the interphase region was reported by Gähde.[10] Although vinylsilanes graft to polyethylene, while methacrylate silanes homopolymerize at the interface, the methacrylate silane was most effective in increasing the crystallinity of polyethylene at the interface. Mechanical properties of kaolin-filled polyethylene seemed to be more dependent on polymer structure at the interface than on chemical grafting.

### 7.3.2. Inhibition of Cure

Fillers are known to have varying degrees of catalytic effect on thermosetting resins that generally inhibit their cure.[11] Among these, alumina trihydrate strongly inhibited the cure of polyester resins, while zirconium silicate inhibited the cure only slightly (Figure 7.8).

Hosaka and Meguro[12] added various ratios of milled E-glass to a room-temperature curing polyester and followed the reaction by infrared spectroscopy. The filler strongly inhibited the polymerization of polyester double bonds with little effect on polymerization of the styrene diluent.

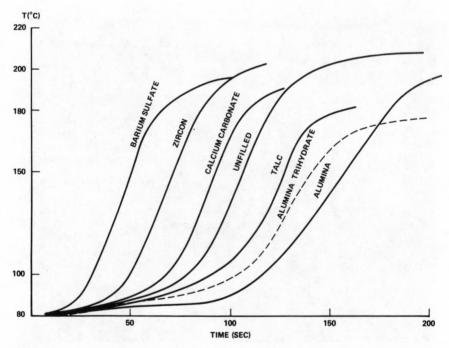

**Figure 7.8.** Polyester exotherms with various fillers (10 g of filler in 20 g of resin in 90°C bath).

Styrene was polymerized with benzyl peroxide in refluxing benzene, with and without the addition of powdered fillers. Examination of the polymer by gel permeation chromatography showed normal molecular-weight distributions at about 26,000 molecular weight. Polymer formed in the presence of zircon contained an additional 2.4% of a very low-molecular-weight fraction, while alumina trihydrate gave 13% low-molecular-weight fraction. Similar observations were made with an azo initiator or with gamma-ray initiation, or with methyl methacrylate monomer and benzoyl peroxide initiation (Table 7.3).

It is evident that filler surfaces inhibit free-radical polymerization by terminating growing chains and not by some catalytic effect on the initiator. Hosaka and Meguro[12] studied charge-transfer phenomena at oxide surfaces and were able to use differences in anion radical formation by metal oxides with strong electron acceptors to estimate the relative electron-donor properties of the oxide surfaces. There is very good agreement between the relative electron-donor properties of oxides as reported by Hosaka with the lowering of polyester exotherms by the same oxides (Table 7.4).

An oxide surface that terminates a free-radical reaction by electron transfer must become an inorganic free radical. Such oxide radicals do not have the necessary activity to initiate new radical chain reactions at low temperatures, but may have adequate activity at higher temperatures. Thus, glass and other fillers may have a severe inhibiting effect on room-temperature-cured polyesters, but cause little difficulty in matched-die molding with high-temperature peroxide initiation.

**Table 7.3.** Solution Polymerization in the Presence of Fillers
(0.1% BPO and 10% monomer in refluxing benzene)

| Monomer | Filler | Polymer (peak MW) | Dimer peak height (% of polymer peak) |
|---------|--------|-------------------|----------------------------------------|
| Styrene | None | 26,700 | 0.0 |
| Styrene | Zircon | 26,600 | 2.4 |
| Styrene | Al-hydrate | 25,800 | 13.0 |
| MMA | None | 47,100 | 0.0 |
| MMA | Al-hydrate | 46,000 | 5.0 |
| *With Luazo-79 initiator* | | | |
| Styrene | None | 31,500 | 0.0 |
| Styrene | Al-hydrate | 32,800 | 10.0 |
| *Gamma rays at 60°C* | | | |
| Styrene | None | 2,700 | 0.0 |
| Styrene | Al-hydrate | 425 | 100 (no high polymer) |

**Table 7.4.** Electron Transfer from Oxide (Hydroxide) Surfaces

| Mineral filler | Relative electron-donor activity[a] | Exotherm lowering by filler in polyester, °C[b] |
|---|---|---|
| MgO | 100 | 24 |
| $Al_2O_3 \cdot 3H_2O$ | 76 | 19 |
| ZnO | 47 | 13 |
| $Al_2O_3$ | 38 | 9 |
| $TiO_2$ | 1.6 | 12 |
| $SiO_2$ | 0.11 | 6 |
| Asbestos (Mg silicate) | — | 25 |
| Talc (Mg silicate) | — | 24 |
| Clay (Al silicate) | — | 19 |
| Wollastonite (Ca silicate) | — | 8 |
| E-glass (mixed silicates) | — | 21 |

[a] H. Hosaka (1974).[12]
[b] 5-g filler in 30-g resin (Plueddemann and Stark, 1973).[11]

Epoxy cures with aliphatic or aromatic amines are also strongly inhibited by glass and other mineral fillers. This inhibition is less serious than with polyesters, since cure can be brought to completion by longer post cures. In contrast, it was observed that cure of epoxies by dicyandiamide at 150° showed little evidence of inhibition by glass.[13] In general, it may be useful to use high-temperature curing conditions with fillers that inhibit the cure too much under moderate conditions.

In both polyesters and epoxies, it was observed that silane treatments of fillers often overcome cure inhibition as measured by cure exotherms (Table 7.5). Silanes that allowed maximum exotherms, that is, were most effective in overcoming surface inhibition, were generally the most effective coupling agents, as indicated by mechanical properties and chemical resist-

**Table 7.5.** Resin Exotherms with Silane-Treated Glass Filler (5-g E-glass in 30-g Resin)[a]

| Organofunctional silane 0.2% on glass | Polyester-$Bz_2O_2$ $\Delta T$, °C | Epoxy-"Z" $\Delta T$, °C |
|---|---|---|
| Untreated | −22 | −17 |
| Epoxy | −20 | −8 |
| Diamine | −15 | −1 |
| Mercaptan | −11 | −8 |
| Cationic styryl | −10 | 0 |
| Phenyl | −7 | −15 |
| Methacrylate ester | −3 | — |

[a] Exotherms in oil bath at 115°C (epoxy) or water bath at 90°C.

ance of composites. A "restrained layer" rather than a "deformable layer" must, therefore, be the best morphology in the interphase region for optimum properties of composites.

## 7.4. Rheology

### 7.4.1. General Concepts

When reinforcing and extending fillers are added to polymers, the resulting material is a complex rheological fluid that is practically impossible to characterize in terms of classic ideal fluids. Silane-treated fillers designed for improved coupling to thermoplastic resins often do have improved processability in compounding and injection molding, but this effect has not been studied systematically or quantitatively.

Filled thermosetting resins are somewhat simpler to study, since the uncured resins generally have low viscosity and resemble paint vehicles in rheology. Paint manufacturers for years have used surface-treated pigments or used surface-active additives to control wet-out, dispersion, thixotropy, and viscosity, and to prevent settling of pigments and fillers. Many commercial surface-active agents are available, and their functions as paint additives have been described,[14] but a comparable study of fillers in liquid thermosetting resins has just begun.[1]

Dispersion of a filler in a liquid polymer involves three steps: (1) Wetting of filler surface; (2) separation of filler particles; and (3) stabilization of the dispersion. Stable dispersion of filler in the final composite is necessary to eliminate filler agglomerates that would act as weak points for chemical or mechanical failure.

After a filler is wet by the vehicle, filler agglomerates are broken up by mechanical shear and the dispersed particles are stabilized by entropic repulsion. Most effective polymer structures for stabilization are linear chains with polar functional groups spaced widely on the chain.

Pretreated fillers may be much easier to disperse in an organic phase because the water layer, which acts as an adhesive to hold filler agglomerates together, is replaced (or covered) by an organofunctional silane.

### 7.4.2. Wetting

Untreated inorganic pigments and fillers have water adsorbed on their surfaces, which hinders their dispersion in organic vehicles. Burrell[15] proposed that all untreated fillers  behave essentially alike with regard to wetting, and require a rather high hydrogen bonding capacity (e.g., medium alcohols or glycol monoethers) in the organic phase for optimum wetting.

The wettability of silane-treated fillers will depend on the nature of the organofunctional group on silicon and on the orientation of the silane on the filler surface. Orientation of cationic silanes on a surface may be controlled by adjusting the pH of the aqueous silane solution. Optimum performance is generally obtained when the silane is applied from a solution pH at the isoelectric point of the filler surface.

Powdered E-glass was treated with aqueous solutions of the cationic silane "H" at several pH's to obtain about 0.1% pickup of silane.

Wet-out time of silane-treated fillers was tested with polyester, epoxy, mineral oil, polyethylene glycol E-600, and water. The polyester was undiluted Rohm and Haas Paraplex® P-13. The epoxy was a 50/50 mix of Dow DER® 330 and XD 7114. Two drops of each liquid were placed in a lightly tamped depression in the treated filler and observed for time of complete soak-in (Table 7.6).

Glass treated from acid dispersions had the fastest wet-out, but was rather hydrophilic. A hydrophobic filler with fair wet-out by organic liquids was obtained when the silane treatment was applied at pH 5.6 with $CO_2$ as acid buffer.

Powdered E-glass treated according to Table 7.6 was used as filler in polyester (P-43) and epoxy (cured with $m$-PDA) castings as shown in Table 7.7. Hydrophilic glass obtained from strongly acidified silane bath gave castings with the poorest wet-strength retention. Alkaline treatments gave poorer wet-out, but very good strengths. Rate of wet-out apparently is not a limiting factor in preparing E-glass composites with liquid thermosetting resins, but hydrophilic residues were especially detrimental in polyester composites.

To study the effect of change in organofunctional silane structure on wet-out of treated glass, microspheres (Potters 3000) were treated with various derivatives of a diaminofunctional silane (silane F of Table 6.1)

**Table 7.6.** Wettability of Powdered E-Glass (10 $\mu m$)
Treated with Aqueous Silane H

| Silane treatment bath | | Wet-out time, min | | | | |
|---|---|---|---|---|---|---|
| Additive | pH | Polyester | Epoxy | Oil | Glycol | $H_2O$ |
| Formic acid | 2.5 | 20 | 3 | 2 | 3 | 1/4 |
| Acetic acid | 3.6 | 60 | 5 | 2 | 6 | 7 |
| $CO_2$ | 5.6 | 180 | 6 | 5 | — | 120 |
| None | 8.2 | 240 | 10 | 3 | 25 | 120 |
| KOH | 12.0 | 240 | 14 | 2 | 25 | 120 |
| No silane | — | 240 | 9 | 6 | 4 | 3 sec |

**Table 7.7.** Flexural Strengths of Powdered E-Glass composites, MPa

| Silane treatment bath | | Polyester | | Epoxy (MPDA) | |
|---|---|---|---|---|---|
| Additive | pH | Dry | Wet | Dry | Wet |
| Formic acid | 2.5 | 136 | 86 | 150 | 121 |
| Acetic acid | 3.6 | 138 | 123 | 143 | 112 |
| $CO_2$ | 5.6 | 139 | 130 | 141 | 118 |
| None | 8.2 | 154 | 126 | 166 | 140 |
| KOH | 12.0 | 146 | 126 | 148 | 118 |
| No silane | — | 112 | 74 | 118 | 47 |

from aqueous baths at pH 4, dried, and tested for wet-out by polyester, epoxy, mineral oil, polyglycol, and water (Table 7.8).

Polar aromatic substituents provided a hydrophobic surface with improved wet-out by polyester. A dodecylbenzyl derivative was not wet-out readily by any of the liquids. This indicates that hydrophobic surfaces are not always organophilic, but that the organic group must have good compatibility with the organic resin for maximum wet-out.

### 7.4.3. Separation and Stabilization of Filler Particles

Adsorbed water acts as an adhesive to cement ultimate particles together. No paint vehicle (or liquid resin) has H-bonds strong enough to break this water bond chemically. If the particles are separated by mechanical shear, they reagglomerate in the absence of something to hold them apart. Once the agglomerate components have been sheared 100–1000 Å apart, a polymer molecule can move in and stabilize the dispersion by entropic repulsion.

**Table 7.8.** Glass Microspheres (Potters® 3000) Treated with Silanes

| R-Group in silane | Wet-out time, min | | | | |
|---|---|---|---|---|---|
| $(CH_3O)_3Si(CH_2)_3NHCH_2CH_2NHR$ | Polyester | Epoxy | Oil | Glycol | $H_2O$ |
| Control (no silane) | 6 | 3 | 3 | 2 | 2 sec |
| H (unmodified silane F) | 20 | 1 | 1 | 3 | 2 sec |
| $-CH_2CH_2COOCH_3$ | 14 | 3 | 2 | 3 | 2 sec |
| $-CH_2C_6H_5$ | 10 | 3 | 2 | 6 | >2 hr |
| $-CH_2C_6H_4CH=CH_2$ | 10 | 3 | 2 | 10 | >2 hr |
| $-CH_2C_6H_3Cl_2$ | 4 | 5 | 1 | 1 | > 2 hr |
| $-CH_2C_6H_4C_{12}H_{25}$ | 70 | 120 | 14 | 180 | > 2 hr |
| $-CH_2C_6H_4OC_6H_5$ | 12 | 3 | 1 | 2 | > 2 hr |

A simple means of determining the relative dispersibility of treated fillers in a liquid resin vehicle is the Daniel flow-point test,[16] which consists of adding a liquid vehicle (15–25% resin in a neutral solvent) from a buret to a known amount of dry pigment while stirring with a spatula. A first endpoint (wet point) occurs when the pigment barely sticks together to form a ball. The second endpoint (flow point) is observed when the mixture becomes fluid enough to flow off the spatula and the last drop falling pulls with it a string that breaks and snaps back. (Highly thixotropic mixes do not give sharp endpoints.)

The wet point measures the amount of liquid necessary to wet the pigment but not disperse it. The flow point measures the amount of liquid necessary to fill the interstices between the pigment particles plus an amount necessary to provide sufficient coating around the pigment particles to deflocculate them.

Different lots of a silane-treated filler often give very different performances in composites. These differences are believed to reflect the degree of uniformity of the silane treatment on the filler. Differences in uniformity of silane coverage should also affect the dispersion of filler in polymer and be observable by the Daniel flow-point test.

Silical filler (Minusil® 5 $\mu$) was treated by dry-blending with 0.3% by weight of the two best coupling agents (silanes D and H of Table 6.1) for polyester resins. A tenfold dilution of the coupling agents with alcohol solvents was tried in order to get more uniform coverage of the filler. The dried, treated fillers were compared in a Daniel flow test, and in polyester castings containing 50% filler.

In the Daniel flow test, 6 g of filler was titrated with a 25% solution of Paraplex® P-43 in styrene. The wet-out point did not vary much among the silane-treated silicas, but the flow point showed significant differences that correlated very well with flexural strengths of filled castings made with the corresponding fillers (Table 7.9). These data suggest that a good coupling agent on a filler can give very poor as well as very good performance in a composite. Complete dispersion of the coupling agent over the filler surface shows itself in low-flow-point titration and in high mechanical performance of composites. It also suggests that a nonpolar silane (D) does not require a polar solvent to aid dispersion on a filler, but that a cationic silane (H) does not disperse well on the filler by dry-blending unless a polar solvent is added as a carrier. Daniel flow tests could serve well as routine control tests on filler treatments.

### 7.4.4. Viscosity

It has been observed often that silane-treated fillers provide lower viscosities in filled resins than do untreated fillers. Addition of silanes as

**Table 7.9.** Evaluation of Silane-Treated Silica (Minusil® 5 $\mu$)

| Solvent of silane treatment | Daniel flow point, ml | Flexural strength of castings, MPa | |
|---|---|---|---|
| | | Dry | Wet (2-hr boil) |
| No silane (control) | 7.0 | 103 | 65 |
| Methacrylate silane D | | | |
| from methanol | 5.9 | 140 | 108 |
| from n-butanol | 5.0 | 169 | 144 |
| No solvent | 3.9 | 172 | 172 |
| Cationic vinylbenzyl silane H | | | |
| from i-propanol | 5.0 | 179 | 124 |
| from Dowanol® EM | 3.7 | 166 | 146 |
| No solvent | 5.3 | 165 | 90 |

integral blends to filler–polymer mixtures often lowers the viscosity markedly,[16] but sometimes raises the viscosity. Different silanes may have opposite effects on viscosities of the mixtures. A minimum viscosity is often desired in order to allow incorporation of as much low-cost filler as possible and still maintain good handling properties. It is desirable, therefore, to be able to predict the effect of a given silane addition on viscosity of filled resins.

Marmo and Mostafa[17] and Fowkes and Mostafa[18] showed that filler–matrix interactions could be predicted from acid–base considerations and could be modified by selecting appropriate solvents. Acidic or basic solvents compete for the polymer and the filler and may assist or prevent adsorption of polymer on filler. The acid–base concept was then applied by Plueddemann and Stark[2] to predict the effect of trace addition of silanes on the viscosity of filled polymer systems.

Fillers, polymers, and silanes may be classified as acid, neutral, or basic. The relative acidity of a filler may be designated as the isoelectric point of its surface in water (Table 4.3). Polymers may be characterized as acids and bases from Drago et al.'s[19] correlation of enthalpies of interaction of organic compounds in neutral solvents. Silanol groups of prehydrolyzed silanes are strongly acidic, neutral alkoxysilanes are neutral to slightly basic, while aminofunctional silanes are basic with the possibility of a zwitterion structure in the hydrolyzed form.

It was found that the effect of organofunctional additives on viscosity of filled resins is strongly influenced by acid–base reactions in filler–additive–resin systems. It is now possible to predict which silane will have optimum effect on viscosity as well as on strength due to "coupling" during resin cure.

1. Neutral polymers required surface-active additives with all fillers for good dispersion. Almost any polar additive will lower viscosity markedly, but acid-functional additives are most effective with basic fillers, and basic additives are recommended with acid fillers.
2. Acid fillers in basic polymers or basic fillers in acid polymers give fairly good dispersions without additives. Lewis acids (titanates and aluminum alkoxides) may be beneficial on acid fillers in basic polymers, but should not be used with basic fillers in acid polymers.
3. Additives may be very helpful in dispersing acid fillers in acid polymers or basic fillers in basic polymers. Cationic silanes or Lewis acids are of most benefit on acid fillers in acid polymers, and may be of some benefit on basic fillers in basic polymers.
4. Neutral silanes (silanes A–D of Table 6.1) modified with catalytic amounts of an amine or a titanate are generally more effective as additives than the pure silane in modifying viscosity. Performance as a coupling agent also is improved by such modification.

Many other variables were observed, but not studied sufficiently to give quantitative results. The viscosity of a given mixture generally increases with increasing humidity. For example, addition of silane H (Table 6.1) to a talc-filled polyester increased the viscosity under conditions of high humidity, but gave marked reduction in viscosity when the filler was predried at 100°C. These observations are illustrated in Table 7.10, showing the viscosity change in filled polyester (Paraplex® P-13, Rohm & Haas) mixes by addition of 0.4% additive based on filler.

It is important, also, to demonstrate that mixtures of silanes and catalysts or other surface-active additives retain (and even enhance) the coupling activity of the silane. A methacrylate-functional silane (D) with

**Table 7.10.** Viscosity Change in Filled Polyester, %

| 0.4% additive/filler | Silica (acid) | Clay (acid) | Talc (weak acid) | $Al(OH)_3$ (weak base) | $CaCO_3$ (base) |
|---|---|---|---|---|---|
| Undecenoic acid | −20 | −3 | −5 | +5 | +87 |
| 1-Hexylamine | −89 | −96 | −47 | +110 | −100 |
| $Ti(Obu)_4$ | −73 | −86 | −34 | +400 | +100 |
| Methacrylate silane D | −19 | −7 | −15 | −12 | −16 |
| Prehydrolyzed methacrylate silane D | −30 | −33 | −4 | −21 | +27 |
| 9/1 $D/Ti(OBu)_4$ | −89 | −59 | −22 | +69 | +72 |
| Cationic silane H | −83 | −92 | −31 | +55 | +29 |

modifiers was compared with a cationic silane (H) and with organic titanates as combined coupling agent/viscosity reducers in silica-filled polyester castings (Table 7.11). Addition of 10% condensation catalyst to silane (D) improved its performance as a coupling agent by facilitating condensation with the silica surface, but gave only minor improvements in viscosity. The cationic silane (H) was most effective in combined viscosity reduction and coupling. Organic titanates were effective in reducing viscosity, but provided no coupling.

## 7.5. Coupling and Adhesion

### 7.5.1. General

Adhesion across the interface in filled polymer composites is essential for improved shear strength, chemical resistance, and retention of electrical properties of composites. Improved impact strength is not a good measure of coupling activity, because treatment of fillers with release agents can impart very high impact strength to composites.

Certain generalizations may be used as guides in selecting a suitable coupling agent for a given system and will be discussed below.

**7.5.1.1. Reactivity.** The organofunctional group of the coupling agent should have maximum reactivity with the resin system under conditions of cure. In comparing a number of unsaturated silanes as coupling agents for

**Table 7.11.** Properties of Polyester (P-43) with Silica Filler
(50% Minusil 5 $\mu$)

| 0.4% additive based on filler | Change in viscosity of mix (%) | Flexural strength of castings, MPa | |
|---|---|---|---|
| | | Dry | Wet (2-hr boil) |
| None | — | 115 | 70 |
| Silane D (Table 6.1) | −2 | 163 | 139 |
| 10% TBT in silane D | −5 | 178 | 152 |
| 10% Hexylamine in silane D | −3 | 184 | 163 |
| TBT | −34 | 106 | 74 |
| TTM-33[a] | −55 | 135 | 72 |
| Cationic silane H | −61 | 184 | 130 |

[a] TTM-33 Kenrich Petrochemicals isopropyltrimethacryltitanate.

glass–polyester resin composites, it was observed[20] that improvement imparted by the silane was directly related to the known reactivity of the unsaturated silanes in copolymerizing with styrene. Several organofunctional silanes are effective coupling agents for epoxy laminates cured at high temperatures with aromatic amines or cyclic anhydrides. Some of these same silanes (e.g., γ-chloropropyltrimethoxy silane) are completely ineffective in epoxy laminates cured with aliphatic amines at room temperature.

**7.5.1.2. Orientation.** Orientation of the silane on the filler (controlled by pH of application) should be such as to allow maximum contact of the functional group with the resin during its cure. Electrokinetic control of orientation of silanes is especially important when cationic-functional silanes are applied to fillers from water. Treatment of silica with cationic silane (H) (Table 6.1) was most sensitive to pH, with a steep improvement in composite properties as the pH of treatment was reduced from 12 to 2. A nonionic silane (D) responded less dramatically to pH of application, but also was the most effective at a pH of 2. In general, cationic silanes are best applied to fillers from aqueous media at a pH correpsonding to the isoelectric point of the surface (IEPS) (cf. Table 4.3).

### 7.5.2. Thermosetting Resins

**7.5.2.1. Styrene-Butadiene (Buton).** Vanderbilt and Clayton[21] first showed that silanes commonly used in fiberglass laminates were also effective in improving strengths of particulate-mineral castings (Figure 7.9). As little as 0.2% vinylsilane converted silica or powdered glass into reinforcing fillers with over 100% improvement in flexural strength. Alumina showed similar improvement but required more silane. Clay, talc, and alumina trihydrate gave more modest improvement,while calcium carbonate and barium sulfate showed no improvement in reinforcement with added silane. Castings with good initial strength also showed exceptional water resistance, with no measurable loss in strength after 7 days in boiling water.

**7.5.2.2. Polyesters.** Data obtained by Sterman and Marsden[22] illustrating the improvements seen in a typical filled polyester casting (Paraplex® P-43) using various fillers and the methacryloxyfunctional silane (D of Table 6.1) are shown in Table 7.12. These data represent integral blend addition of silane D and provide a good indication of the silane response to be expected from a given mineral filler in many thermoset composites. The performance of various silanes on glass in reinforced polyester composites are summarized in Tables 6.1–6.4.

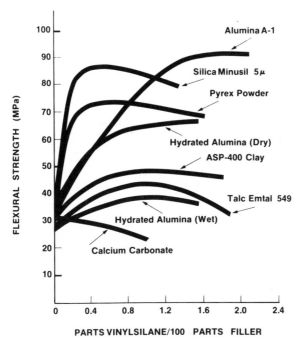

**Figure 7.9.** Strengths of particulate–mineral castings as a function of flexural strength vs. parts vinylsilane/100 parts filler.

**Table 7.12.** Performance of Polyester Resin Composites with Various Fillers and Silane D (Table 6.1)

| Filler | Silane % D added | Flexural strength, MPa | | % Improvement over nonsilane control | |
| | | Initial | 8-hr boil | Initial | 8-hr boil |
|---|---|---|---|---|---|
| Quartz | 1.30 | 138 | 103 | 75 | 130 |
| Wollastonite | 0.80 | 124 | 103 | 38 | 100 |
| Aluminum silicate | 0.30 | 103 | 97 | 25 | 100 |
| Clay-chopped glass | 2.25 | 159 | 117 | 29 | 25–69 |
| Calcined clay | 2.20 | 103 | 97 | 15 | 27 |
| Hydrous clay | 2.20 | 83 | 62 | 25 | 22 |
| Hydrated alumina | 1.00 | 76 | 55 | 28 | 15 |
| Calcium carbonate | 2.00 | 83 | 62 | 20 | Nil |

**7.5.2.3.   Epoxies.** Silanes are used in mineral-filled epoxy systems to provide electrical insulation and low-loss properties after extensive exposure to water.[23] The electrical property protection afforded by the use of silane coupling agents in quartz and Wollastonite-filled epoxy composites is shown in Table 7.13. The addition of only 0.3 phr of silane (either D or E of Table 6.1) stabilizes the dielectric constant at about 2.4, similar to that of unfilled epoxy.

Mineral-filled molding compounds with improved mechanical strength and electrical properties result from silane pretreatment on silica and silicate fillers, but not from pretreatment on calcium carbonate fillers. Addition of silanes to molding compounds as an integral additive may make it difficult to get easy release from molds.

Fiberglass-reinforced epoxy composites are described in Figure 6.1 and Tables 6.6–6.9.

**7.5.2.4.   Phenolic Resins.** Phenolic composites with high loadings of three fillers—glass spheres, $Al_2O_2$ abrasive grit for foundary wheels, and foundary sand for shell molding—were prepared with silanes E or C added to the resin at low levels.[24] Addition of a trace of silane raised the initial strength to some extent, but showed the most dramatic improvement in wet-strength retention (Table 7.14).

Two silanes are compared in glass cloth-reinforced phenolic composites in Table 6.12. In general, silanes that are effective in glass fiber-reinforced composites are also preferred in particulate mineral-filled composites.

**7.5.2.5.   Other Thermosetting Resins.** Silanes A, D, and H (of Table 6.1) are recommended as coupling agents for other unsaturated resins based on diallylphthalate, acrylic monomers, and cross-linkable polyolefins. Vinyl silanes (A) are used in mineral-filled cross-linkable polyethylene for improved retention of electrical properties in cable coverings. The cationic silane H is preferred for coupling fillers to acrylic-modified epoxies (vinyl resins). Silanes C, E, F, or H are recommended coupling agents for melamine, furan, and polyimide resins (see Table 6.15).

### 7.5.3.  Thermoplastic Resins

**7.5.3.1.   General.** Silane treatments on particulate mineral fillers may provide important improvements in rheology of filled thermoplastics and in protecting the filler against mechanical damage during high-shear operations such as mixing, extruding, and injection molding. Such mechanical improvements may be as important as any chemical coupling across the interface in filled systems. Reasonable explanations have been

**Table 7.13.** Epoxy Resin Composites: Electrical Properties Using Quartz and Wollastonite Fillers[a] and Silanes C or E

| Silane added[b] | Quartz | | | | Wollastonite | | | |
| | Dielectric constant | | Dissipation factor | | Dielectric constant | | Dissipation factor | |
| | Initial | 72-hr boil | Initial | 72-hr boil | Initial | 72-hr boil | Initial | 8-hr boil |
|---|---|---|---|---|---|---|---|---|
| Unfilled resin | 3.44 | 3.43 | 0.007 | 0.005 | 3.44 | 3.43 | 0.007 | 0.005 |
| Filled, no silane | 3.39 | 14.60 | 0.017 | 0.305 | 3.48 | 22.10 | 0.009 | 0.238 |
| Filled + C | 3.40 | 3.44 | 0.016 | 0.024 | 3.30 | 3.32 | 0.014 | 0.016 |
| Filled + E | 3.46 | 3.47 | 0.013 | 0.023 | 3.48 | 3.55 | 0.017 | 0.028 |

[a] Composition (parts by weight): ERL-2274 (Union Carbide), methylnadic anhydride, 80; benzyldimethylamine, 0.5; filler, 180. Electrical measurements at 1000 Hz per ASTM D-150. Cured 16 hr at 120°C plus 1 hr at 180°C.
[b] Silane mixed with resin–hardener-catalyst at about 0.3 phr.

**Table 7.14.** Phenolic Resin Composites—Aluminum Oxide,
Glass Spheres, and Sand Fillers, Silanes E or C (Table 6.1)

| | Aluminum oxide[a] | | Glass spheres[b] | | Sand shell mold | |
|---|---|---|---|---|---|---|
| | | | | | Core tensile strength, MPa | |
| | Flexural strength, MPa | | Tensile strength, MPa | | | |
| Silane | Dry | Wet[c] | Dry | Wet[d] | Dry | Wet[e] |
| Filled, no silane | 35 | 14 | 2.6 | Nil | 2.2 | 0.1 |
| Filled + E | 41 | 32 | 2.7 | 1.9 | 3.5 | 2.8 |
| Filled + C | — | — | 2.2 | 0.9 | — | — |

[a] Composite contained 77 wt% $Al_2O_3$ pretreated with silane plus wetting oil (60/40 furfural and cresol). Phenolic resin mixture as follows: BRNA-5345 (Union Carbide), 124 parts; cryolite, 3.9 parts; hexamethylenetetramine, 10 phr.
[b] Composite contained 97 wt% glass spheres (Cataphote Co.) and 3 wt% Pliophen 5671 (Reichhold Chem.). Cured 7 min at 450°F.
[c] Conditioned 10 days in water at room temperature.
[d] Condidtioned 16 hr in water at 50°C.
[e] Conditioned 16 hr at RT and 100% RH.

proposed for coupling through organofunctional silanes even to chemically nonreactive thermoplastics (see Section 5.3.3).

Glassy amorphous thermoplastics have sufficient physical compatibility with a number of organofunctional silanes to allow good bonding to silane-treated surfaces.[25] Crystalline thermoplastics are generally incompatible with nonreactive organofunctional silanes, but may be bonded by grafting reactions of the treated surface with the polymer. Suitable silanes for grafting may have vinyl, peroxy,[26] or axido[27] functional groups. Reactive silanes with cationic vinylbenzyl H or methacrylate D apparently do not graft very well to thermoplastics, but may homopolymerize to form pseudo-interpenetrating polymer networks with the thermoplastic at the interface.[28]

**7.5.3.2. Polyolefins.** Filled polyethylene to be extruded as cable covering is modified with silanes to improve wet electrical properties of the composite (Table 7.15). Clay, Wollastonite, and quartz-filled composites clearly show the improvement through the addition of methacrylate-functional (D) and epoxy-functional (C) silanes (of Table 6.1).

Fillers treated with the vinylbenzyl cationic silane H gave very good dispersion in polyolefins by simple mixing in the screw of an injection molder. Coupling appears to be through an interpenetrating polymer network at the interface, and is improved by adding a trace of dicumyl peroxide to the silane and powdered chloroparaffin to the polyolefin.[28] The

**Table 7.15.** Electrical Properties of Mineral-Filled Polyethylene[a]

| Silane added[b] | Clay | | | | Wollastonite | | | | Quartz | | | |
|---|---|---|---|---|---|---|---|---|---|---|---|---|
| | Dielectric constant | | Dissipation factor | | Dielectric constant | | Dissipation factor | | Dielectric constant | | Dissipation factor | |
| | Initial | Wet | Initial | Wet | Dry | Wet | Dry | Wet | Dry | Wet | Dry | Wet |
| Filled, no silane | 2.7 | 3.0 | 0.003 | 0.082 | 2.8 | 4.2 | 0.009 | 0.147 | 2.7 | 5.2 | 0.029 | 0.228 |
| Filled + D | 2.7 | 2.7 | 0.002 | 0.003 | 2.8 | 2.9 | 0.007 | 0.014 | — | 2.5 | 0.010 | 0.012 |
| Filled + C | 2.7 | 2.7 | 0.002 | 0.005 | 2.8 | 2.9 | 0.007 | 0.013 | — | — | — | — |

[a] Resin, filler, and silane mixed in a two-roll mill at 240–250°F. Composition: Polyethylene resin DYNH (Union Carbide), low density, 50 wt%. Polyethylene resin DMD-7000 (Union Carbide), high density, 50 wt%; filler 50 wt%. Electrical properties measure at 1000 Hz per ASTM D-150.
[b] Silane added to clay at 2 phf; to Wollastonite at 0.8 phf; and to quartz at 0.3 phf.

**Table 7.16.** Polyolefins with 35% Mica[a] Filler

| Coupling system with filler | H.D. Polyethylene[b] | | Polypropylene[c] | |
|---|---|---|---|---|
| | Flexural strength, MPa | Tensile strength, MPa | Flexural strength, MPa | Tensile strength, MPa |
| None | — | — | 59 | 28 |
| Silane H (Table 6.1) | 34 | 19 | — | — |
| H + P[d] | 36 | 21 | 63 | 29 |
| H + P + CP[e] | 40 | 25 | 75 | 38 |

[a] Suzerite 325-H (Martin Marietta Co.).
[b] XP-3574.30 (Dow).
[c] Profax 6223 (Hercules).
[d] 1% dicumyl peroxide added to silane.
[e] 1% Chlorez® (Dover Chem. Co.) and 0.05% MgO added to PP.

effects of silane, peroxide, and chlorinated paraffin are cumulative in coupling minerals to polyolefins as shown in data on mica-filled polyethylene and polypropylene (Table 7.16).

Improved coupling in mineral-filled polyolefins generally raises the heat-distortion temperature of the composite, but sometimes causes a loss in impact strength (cf. Section 1.3.7). The combination of silane H (Table 6.1) with peroxide and chlorinated paraffin raised the heat distortion temperature of Wollastonite-filled polypropylene as expected, but without any decrease in impact strength (Table 7.17).

**7.5.3.3. Engineering Thermoplastics.** Organofunctional silanes that are recommended for coupling to epoxy resins seem to perform rather well in mineral-filled nylon. Nylon 6/6 was injection-molded with 35% mineral

**Table 7.17.** Polypropylene with 35% Wollastonite[a] Filler

| Coupling system with filler | Flexural strength MPa | Tensile strength MPa | HDT, °C at 1.8 MPa (264 psi) | Falling dart impact (J) |
|---|---|---|---|---|
| No filler | 60 | 33 | 56 | 2.8 |
| No treatment | 58 | 28 | 90 | 6.7 |
| Silane H (Table 6.1) | 60 | 29 | — | 6.7 |
| H + P[b] | 65 | 33 | 118 | — |
| H + P + CP[c] | 68 | 33 | 118 | 6.7 |

[a] Nyad-G (Nyco Div.) in Profax 6223 (Hercules).
[b] 1% dicumyl peroxide added to silane.
[c] 1% Chlorez® 700 (Dover Chem.) and 0.05% MgO added to PP.

fillers treated with typical silanes.[29] All composites with treated filler gave about 30% improvement in dry strength and 40–50% improvement in wet strength compared to untreated fillers (Table 7.18).

Injection-moldable thermoplastics with discontinuous glass fiber reinforcement are available commercially. The glass fibers are sized with proprietary formulations containing organofunctional silanes. Aminoalkyl-silanes are used with a surprising number of thermoplastics, although epoxy-, vinyl-, methacrylate-, or other organofunctional silanes may be used. Performance of typical glass-reinforced thermoplastics was summarized by Arkles,[30] as shown in Table 7.19.

The degree of improvement in mechanical properties imparted by silanes to filled injection-molded thermoplastics is generally less marked than the improvement observed in similar compression-molded systems.[1] For example, the "best" improvement imparted by silanes in compression-molded thermoplastics is compared with comparable improvement imparted by silanes in filled injection-molded thermoplastics in Table 7.20. It appears that optimum coupling and reinforcement are not being obtained in filled injection-molded thermoplastics. This failure may be due to fracture in the filler or reinforcement under high shear of injection molding, but it may also be due to a failure in the interphase region (coupling agent interaction with reinforcement and resin) under high shear.

Numerous efforts have been made to improve the performance of silane coupling agents in filled injection-molded composites. It was proposed that polymeric silanes, or silane-modified polymer film formers, might distribute the stresses more uniformly across the interface. Polymeric silanes showed a significant advantage over monomeric silanes as filler treatment in kaolin-filled compression-molded EPDM elastomer, but they showed little advantage in 30% fiberglass-filled nylon 6/6 in composites formed by injection molding.[31]

Shear rates during injection molding are up to 1000 times as great as they are in compression molding.[32] Any structure built up in the interphase region is torn from the filler during this period of high shear. What is needed is an interphase region that is highly fluid during the molding operation, but sets to a tough polymer when cooled.

### 7.5.3.4. Ionomer Bonds Across the Interface.

The property desired in the interphase region of thermoplastic composites fits the description of commercial ionomers.[33] Commercial ionomers are thermoplastic polymers modified with up to 10% acrylic (or methacrylic) acid and neutralized about 30% with $Na^+$ or $Zn^{2+}$. Compared to the parent polymer, ionomers have lower coefficients of thermal expansion, greater toughness, and greater tear resistance. Ionomers are believed to consist of ion clusters dispersed in a

**Table 7.18.** Properties of 35% Mineral-Filled Nylon 6/6

| Silane treatment on mineral (Table 6.1) | Glass microspheres[a] | | | | Novacite silica[b] | | | |
| --- | --- | --- | --- | --- | --- | --- | --- | --- |
| | Flexural strength MPa | | Tensile strength MPa | | Flexural strength MPa | | Tensile strength MPa | |
| | Dry | Wet | Dry | Wet | Dry | Wet | Dry | Wet |
| None | 10.5 | 6.5 | 5.2 | 4.2 | 11.1 | 6.9 | 6.0 | 4.3 |
| Amino F | 14.7 | 8.8 | 8.2 | 6.3 | 14.6 | 9.6 | 8.4 | 6.8 |
| Chloroalkyl B | 14.7 | — | 8.3 | — | 14.8 | — | 8.2 | — |
| Cationic styryl H | 14.7 | 9.4 | 8.3 | 7.0 | 14.5 | 9.7 | 8.5 | 6.6 |

[a] Potters 3000.
[b] Malvern Minerals 207 A.
[c] Wet: after 16 hr in water at 50°C.

**Table 7.19.** Properties of 30% Fiber-Glass-Reinforced
Thermoplastic Composites

| Polymer | Tensile yield strength, MPa | Flexural modulus, GPa | Notched impact (J) | HDT, °C at 1.8 MPa (264 psi) |
|---|---|---|---|---|
| ABS | 10.0 | 7.59 | 0.16 | 104 |
| Acetal | 13.4 | 9.66 | 0.20 | 163 |
| Nylon | 15.8 | 8.28 | 0.26 | 216 |
| Polycarbonate | 12.8 | 8.28 | 4.11 | 149 |
| PPO-base resin | 14.5 | 8.96 | 0.19 | 154 |
| Polysulfone | 12.4 | 8.28 | 0.20 | 185 |
| PP sulfide | 13.8 | 11.00 | 0.16 | 260 |
| Polystyrene | 9.3 | 8.96 | 0.11 | 102 |
| SAN | 12.0 | 10.30 | 0.16 | 102 |
| Polyester (PBT) | 13.4 | 9.66 | 0.19 | 221 |
| PVC | 9.0 | 0.59 | 0.13 | 68 |

neutral thermoplastic matrix. Addition of cations, by coordinating with carboxyl groups, increases the tendency for clustering. From practical experience with commercial ionomers it has been observed that optimum properties are achieved at about 33% neutralization of available acid groups. At high degrees of neutralization the melt viscosity increases, as does the sensitivity to water. There is relatively little effect of changing ions from $Na^+$ to $Zn^{2+}$. At high temperatures and under high shear, the ionomer clusters are fluid, but they act as cross-links when cooled to room temperature.

**Table 7.20.** Improvement Imparted by Silanes in
Glass-Thermoplastic Composites

| Polymer–silane system (Table 6.1) | Flexural strength improvement, % | | | |
|---|---|---|---|---|
| | Compression-molded | | Injection-molded | |
| | Dry | Wet | Dry | Wet |
| Nylon-aminosilane F | 55 | 115 | 40 | 36 |
| Nylon-cationic silane H | 85 | 133 | 40 | 45 |
| PBT-aminosilane F | 21 | — | 23 | 24 |
| PBT-cationic silane H | 60 | 47 | 28 | 11 |
| Polypropylene silane F | 8 | 18 | 7 | 10 |
| Polypropylene silane H | 86 | 89 | 16 | 16 |

Ionic bonds in themselves do not contribute water resistance across an interface. Carboxy-modified polymers have very good dry adhesion to glass but lose the adhesion in water. Commercial ionomers also develop strong dry bonds when melted against glass and metals, but lose adhesion in water.

Acid-functional polymers will bond directly to aminofunctional silanes on surfaces, but the covalent amide bonds formed are torn from the surface under conditions of high shear. In the preferred embodiment of the "ionomer bonding" concept, an acid-functional silane is bonded to the reinforcement through stable siloxane bonds. The acid-functional silane bonds to an acid-functional polymer through an ionomer bond that is mobile at conditions of molding, but sets to a tough water-resistant bond at room temperature. When the matrix polymer contains no acid groups, an acid-functional polymer that is compatible with the matrix may be included in formulating the ion-modified silane primer. A compatible acid-functional polymer may also be added in small proportion to the matrix polymer and allowed to react with the treated reinforcement during the molding operation. Many acid-functional polymers are available commercially and are recommended as additives to thermoplastic resins for improved adhesion to commercial (silane-treated) fiberglass. These same systems would give "ionomer-bonded" composites if the glass were treated with salts of acid-functional silanes.

A convenient method of providing acid-functional silane coupling agents is to mix an aminoorganofunctional silane with excess of a dibasic acid such that amic acids are formed during drying at about 175°C. Acids such as orthophthalic, maleic, or succinic that cyclize readily to cyclic imides are less effective than noncyclizing acids such as fumaric, adipic, and isophthalic. The reaction product of a diamine-functional silane (Z-6020) with a 10% excess of isophthalic acid is especially promising because it seems to have very good heat stability. The mixture is soluble in methanol or water and is compatible with sodium and zinc ions.[34]

Nylon 6/6 was pressed at 250°C against primed glass and compared for retention of adhesion in 95°C water (Table 7.21).

Acid-functional film formers for use in ionomer bonding to polyolefins may be formulated directly into the primer (e.g., Primacor® 4983 emulsion, Dow Chemical Co.) or added as a separate phase. It is important that the acid-functional polymer be compatible with the neutral polyolefin. Glass microscope slides were coated with a primer comprising 1 mol silane F (Z-6020), 1.1 mol isophthalic acid and 0.8 mol of zinc acetate in water. The primer was dried 15 min at 175°C and then coated with a thin layer of acid-modified polyolefin and used as substrate for adhesion of unmodified polyolefin fused at 250°C. Water resistance of the bond was good when the film-former was compatible with the polyolefin (Table 7.22).

**Table 7.21.** Bonding Nylon 6/6 to Glass Through Ionomer Primers
(press nylon on primed glass at 250°C)

| Ratio of ion of free COOH in nylon primer | Nylon adhesion to glass in 95°C water | | |
|---|---|---|---|
| | 2 hr | 4 hr | 6 hr |
| No primer | Poor | | |
| Primer with no cations | Excellent | Good | Poor |
| 0.1 Zn$^{2+}$ | Excellent | Excellent | Fair |
| 0.2 Zn$^{2+}$ | Excellent | Excellent | Excellent |
| 0.3 Zn$^{2+}$ | Excellent | Excellent | Good |
| 0.5 Zn$^{2+}$ | Excellent | Good | Poor |
| 0.1 Na$^+$ | Excellent | Good | Poor |
| 0.3 Na$^+$ | Excellent | Excellent | Good |

### 7.5.4. Elastomers

Particulate mineral fillers have been used in organic rubbers for many years, but their total application has been small compared to that of carbon black. Low-cost minerals are not such effective reinforcing fillers as carbon black, and high-surface silicas that are good reinforcing fillers are more costly than carbon black. Only in silicone rubbers have costly high-surface silicas become the standard reinforcement.

Much effort has been expended in upgrading low-cost mineral fillers by treating them with reactive silanes in order to obtain a cost-effective replacement for carbon black. Although much progress has been made in formulating rubbers with silane-treated mineral fillers, it is believed that significant improvements can still be made in present systems. As the pressure continues on energy and carbon black costs, there will be greater incentive to develop silane-modified mineral fillers as replacements for carbon black.

**Table 7.22.** Bonding Polyolefins to Primed Glass

| Carboxylated polyolefin interlayer[a] | Polyolefin top layer | Adhesion after 1 day in water at room temperature |
|---|---|---|
| Primacor® 4983 PE | HDPE | Excellent |
| | HDPP | Nil |
| Polybond® 1001 PP | HDPE | Poor (fail between polymers) |
| | HDPP | Excellent |
| Polybond® 1016 co(PP/20% PE) | HDPE | Excellent |
| | HDPP | Excellent |

[a] Primacor® = product of Dow Chemical Co (Midland, MI); Polybond® = product of BP Performance Polymers, Inc. (Hackettstown, NJ).

Silane treatments on fillers are known to be very effective in improving dispersion in hydrocarbon polymers (see Section 7.4). The treatment also may affect scorch and cure times of sulfur-vulcanized rubbers. It was anticipated that performance of silane-treated fillers in elastomers should be related to the degree of adhesion imparted by silanes between rubber and filler. Peel adhesion of rubbers vulcanized against silane-primed glass microscope slides was correlated with the properties of clay-filled rubbers with the same silanes on the filler.[35]

A hard kaolin clay (Suprex®) was tested in three rubber formulations as shown in Table 7.23. Silane-treated clay was dispersed in the rubbers on the mill before adding the remaining ingredients. The vulcanized compounds were evaluated for selected mechanical properties by standard ASTM methods and correlated with microscope-slide adhesion tests (Table 7.24).

It appears that reaction of organofunctional silanes with rubber types is specific enough that peel adhesion from glass correlates fairly well with mechanical properties obtained with the same silanes on clay filler. Increased adhesion, as indicated by peel tests, consistently improves certain mechanical properties in all rubbers, has little effect on some properties, and varies with the type of rubber in other properties. Each trend in properties is accentuated by increasing the level of silane treatment from 0.5% to 1.0%.

Increased adhesion is accompanied by the following property changes:

1. 300% modulus—increase;

**Table 7.23.** Clay-Filled Rubber Formulations

| Formulation | Natural (SMR-5) | SBR (1502) | Nitrile (FRN-502) |
|---|---|---|---|
| Rubber | 100 | 100 | 100 |
| Clay | 50 | 60 | 70 |
| Zinc oxide | 5 | 4 | 5 |
| Stearic acid | 1.5 | 2 | 2 |
| Sulphur | 2.5 | 1.4 | 0.2 |
| Plastogen® | 5.0 | — | — |
| Sunolite®-240 | 1.5 | — | 1.0 |
| Thermoflex®-A | 1.0 | — | — |
| TMTD | 0.5 | — | 3.0 |
| Santocure®-NS | 1.0 | — | — |
| Circolite® oil | — | 10.0 | — |
| PBNA | — | 1.0 | 1.0 |
| Flexamine®-G | — | 1.0 | — |
| MBT | — | 0.8 | — |
| Di-o-tolylguanidine | — | 0.3 | — |
| DOP | — | — | 10.0 |

**Table 7.24.** Mechanical Properties of Clay-Filled Rubber
(clay treated with 1% aqueous silane)

| Silane on clay | None | Chloropropyl B | Mercaptopropyl G | Amine F |
|---|---|---|---|---|
| | | Natural rubber | | |
| Peel adhesion, N/cm | Nil | 1.1 | 3.5 | 7.7 |
| 300% Modulus, MPa | 7.2 | 6.8 | 10.2 | 11.4 |
| Tensile strength, MPa | 23.6 | 25.4 | 26.8 | 27.1 |
| Elongation, % | 585 | 585 | 540 | 520 |
| Tension set, % | 43 | 43 | 41 | 38 |
| Compression set, % | 8 | 7 | 6 | 6 |
| Bashore rebound, % | 59 | 53 | 64 | 67 |
| Tear strength, ppi | N/cm | 242 | 245 | 207 |
| | | SBR 1502 | | |
| Peel adhesion, N/cm | Nil | 0.2 | 2.1 | 11.0 |
| 300% Modulus, MPa | 2.0 | 2.0 | 2.8 | 2.6 |
| Tensile strength, MPa | 7.7 | 9.5 | 10.4 | 11.9 |
| Elongation, % | 925 | 1015 | 885 | 975 |
| Tension set, % | 37 | 41 | 35 | 38 |
| Compression set, % | 13 | 11 | 10 | 9 |
| Bashore rebound, % | 46 | 47 | 48 | 48 |
| Tear strength, N/cm | 247 | 262 | 270 | 275 |
| Flex, JIS $10^{-2}$ | 337 | 567 | 259 | 520 |
| Abrasion resistance | 100 | 130 | 155 | 140 |
| Heat build-up, °C | $91^a$ | $103^a$ | $90^a$ | 97 |
| | | Nitrile rubber | | |
| Peel adhesion, N/cm | Nil | 0.9 | 2.1 | 14.0 |
| 300% Modulus, MPa | 8.5 | 9.3 | 12.1 | 14.7 |
| Tensile strength, MPa | 21.7 | 22.4 | 24.3 | 24.1 |
| Elongation, % | 650 | 665 | 665 | 545 |
| Tension set, % | 31 | 27 | 25 | 21 |
| Compression set, % | 7 | 7 | 6 | 4 |
| Bashore rebound, % | 24 | 23 | 23 | 24 |
| Tear strength, N/cm | 352 | 308 | 293 | 268 |
| Flex, JIS $10^{-2}$ | 49 | 102 | 104 | 36 |
| Abrasion resistance | 100 | 122 | 141 | 166 |
| Heat build-up, °C | $61^a$ | $98^a$ | $69^a$ | 55 |

[a] Sample blew out before completion of heat buildup test.

2. tensile strength—increase;
3. abrasion resistance—increase;
4. tension set—decrease;
5. compression set—decrease;
6. heat build-up—decrease;

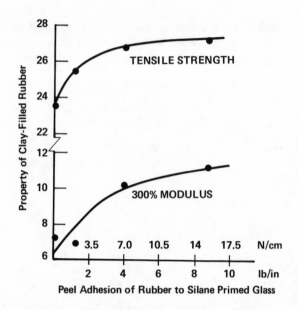

**Figure 7.10.** Relationship of adhesion to rubber reinforcement in clay-filled natural rubber.

7. Shore A hardness—little effect;
8. Bashore rebound—little effect;
9. Mooney viscosity— + natural and nitrile, −SBR;
10. scorch and cure time— + natural and nitrile, −SBR;
11. elongation— + SBR, −natural and nitrile;
12. tear strength— + natural and SBR, −nitrile;
13. flex resistance— + SBR and nitrile, −natural;
14. heat aging—better natural, poorer SBR and nitrile.

The relationship of adhesion to tensile strength and 300% modulus of clay-filled natural rubber is shown in Figure 7.10. This work has not been continued to its logical conclusion of testing silanes on fillers that give perfect adhesion (cohesive failure) of rubber to glass.

## 7.6. Methods of Applying Silanes

### 7.6.1. General

A neutral methacrylate ester-functional silane (D of Table 6.1) and a cationic vinylbenzylfunctional silane H represent two general classes of silanes commonly used as coupling agents: (a) neutral, volatile silanes that

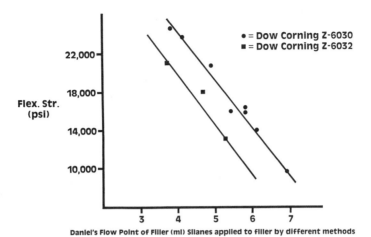

**Figure 7.11.** Daniel's flow point of filler (ml) silanes applied to filler by different methods.

may benefit from added catalysts, and (b) cationic silanes with low volatility that are self-catalyzed for reaction with mineral surfaces, but may require solvent to aid in dispersing onto dry fillers. Both materials are recognized as very effective coupling agents for mineral fillers and reinforcements in unsaturated polyesters.

For any type of treatment with a given silane, it is generally observed that optimum treatment, as shown by reduced viscosity of a polyester mix or by reduced readings in the Daniel's flow test, correlates well with performance of that filler in a polyester composite as shown by flexural strength and chemical resistance (Figure 7.11).

### 7.6.2. Integral Blends

Particulate fillers in liquid resins are most readily modified with coupling agents by adding silanes as integral blends to the mixture. Particulate fillers may also be treated in situ when milled with elastomers by integral blending. In this case, the silane should be added to the filled elastomer before vulcanizing ingredients are added on the rolls. Integral addition of silanes to filled polymer systems depends on adsorption of the coupling agent on the filler during the mixing operation.

Cationic silane H is much more reactive with fillers and gives minimum viscosity and maximum performance with only short contact. Silane D is much slower in hydrolyzing on a filler surface and may require an added catalyst for rapid viscosity effects. Even though it imparts little decrease in viscosity of silica-filled polyesters at room temperature, it still performs fairly well as a coupling agent in the cured composite. Longer contact times and elevated temperature of the cure apparently allow the alkoxysilane to hydrolyze and condense with the hydrated silica surface. A small proportion of silane H mixed with silane D acts as a catalyst for condensation of silane D with the filler and also assists it in bonding to polyester resins (see Table 6.3).

### 7.6.3.  Treatment of Filler from Organic Solvents

The simplest, and most effective, laboratory treatment of particulate fillers is to mix the filler with a dilute solution of silane in an organic solvent, filter, and dry. Silica gel and controlled-pore glass were treated by Waddel, et al.[36] with a solution of a diaminofunctional silane (F of Table 6.1) in refluxing toluene. Treatment from a dry solvent tends to deposita more nearly monomolecular layer of silane than can be obtained from water. Finely divided particulates do not cake up upon drying as they do when treated from water.

### 7.6.4.  Treatment of Fillers from Water

Water dispersions of silanes (and other ingredients) are almost invariably used in treating glass fibers. Precautions must be taken to ensure both stability and proper orientation of silane on the glass surface (see Chapter 3).

Aqueous treatments are very effective on coarse fillers such as glass microbeads and glass fibers that may be dried readily without caking. Finely divided fillers may also be treated from water and recovered by spray-drying to prevent clumping into hard agrregates. This may not be feasible for economic reasons unless filler processing already involves an aqueous treatment where silanes could be included without adding another step to the process.

### 7.6.5.  Treatment of Fillers by Dry-Blending

Silanes may also be dry-blended with fillers at room temperature or at elevated temperature. A small amount of suitable solvent may be beneficial in aiding dispersion of a trace of silane over a large surface area of filler. Neutral, volatile silanes such as vinyltrimethoxysilane or chloropropyltrimethoxysilane disperse most readily. Triethoxy- or tripropoxysilanes migrate readily over the filler, but are much slower in their hydrolysis and reaction with the filler surface. It has been observed occasionally that old samples of treated filler, after standing in closed containers for 6 months, give lower viscosities in polymer mixes, or better performance in composites, than freshly treated filler. This suggests that complete dispersion of silane on the filler may take more than several days. Normal shipping and storage times would be sufficient to allow optimum dispersion of silane on commercially treated fillers.

High-molecular-weight aminofunctional silanes such as H do not diffuse readily when dry-blended with fillers ar room temperature. The tiny proportion of silane needed to treat a typical filler tends to concentrate in localized clumps and is slow in diffusing to the total filler. Dilution of the silane with about 10 times its volume of a medium alcohol or a glycol ether before mixing with filler seems to be rather effective in obtaining dispersion throughout the filler. The treated filler must then be dried before it is used in polymer mixes.

Simple dry-blending of alkoxysilane coupling agents with fillers at room temperature may not get optimum coverage of the filler surface because of two conflicting requirements: dispersion and condensation. Less reactive nonpolar silanes disperse best on the filler, but may not condense sufficiently with the hydrated surface to provide optimum dispersion of filler in the resin or coupling in the composite. Addition of a catalyst (aminofunction silane) or use of elevated temperature may aid in surface treatment. Cationic silanes such as silane H are very reactive with hydrated filler surfaces, even at room temperature, but they may condense too rapidly and not treat the entire surface. Addition of a carrier liquid such as the glycol ether alcohols may be necessary to allow such silanes to completely cover the filler surface.

Simple viscosity tests on mixes of fillers and liquid resin, or Daniel flow-point tests, should be useful screening devices to determine the relative effectiveness of different silane treatments of fillers.

## References

1. E. P. Plueddemann. In *Additives for Plastics*, Vol. I, R. B. Seymour, Ed., p. 49 Academic, New York (1978).
2. E. P. Plueddemann and G. L. Stark. In *Additives for Plastics*, Vol. II, R. B. Seymour, Ed., p. 49 Academic, New York (1978).
3. A. T. DiBenedetto and D. A. Scola. *J. Colloid Interface Sci.* **74**(1), 150 (1980).
4. R. Holtmann. *Kunststoffe* **55**(12), 903 (1965).
5. D. J. Vaughan. Private communication (1980).
6. E. P. Plueddemann. SPI, 29th Ann. Tech. Conf. Reinf. Plast. 24-A (1974).
7. D. J. Vaughan and J. W. Sanders. SPI, 29th Ann. Tech. Conf. Reinf. Plast. 13-A (1974).
8. W. D. Bascom. *Adv. Chem. Ser.* **87**, 38 (1968).
9. F. J. Kahn. *Appl. Phys. Lett.* **22**(8), 386 (1973).
10. J. Gähde. *Plaste u Kautschuk* **22**(8) 626 (1975).
11. E. P. Plueddemann and G. L. Stark. SPI, 28th Ann. Tech. Conf. Reinf. Plast. 21-E (1973).

12. H. Hosaka and K. Meguro. *Prog. Org. Coat.* **2**, 315 (1973–1974).
13. E. P. Plueddemann. World Electrotech. Conf. (Moscow), Sect. 3-A, No. 41 (1977).
14. E. Singer. In *Treatise on Coatings*, R. R. Meyers and J. S. Long, Eds., Vol. 3, Pt. 1, Chap. 1, Marcel Dekker, New York (1975).
15. H. Burrell. *Proc. Amer. Chem. Soc. Div. Coatings Plast. Chem.* **35**(2) 18, 25 (1975).
16. E. P. Plueddemann and G. L. Stark. *Mod. Plast.* **54**, 102 (1977).
17. M. J. Marmo and M. A. Mostafa et al. *Ind. Eng. Chem., Prod. Res. Devel.* **15**(3), 206 (1976).
18. F. M. Fowkes and M. S. Mostafa. *Amer. Chem. Soc., Div. Org. Coat. Plast. Chem., Preprints* **37**(1) 142 (1977).
19. R. S. Drago, L. B. Parr, and C. S. Chamberlain. *J. Am. Chem. Soc.* **99**, 3203 (1977).
20. E. P. Plueddemann, H. L. Clark, L. E. Nelson, and K. R. Hoffman. *Mod. Plast.* **39**(8), 135 (1962).
21. B. M. Vanderbilt and R. E. Clayton. *Ind. Eng. Chem., Prod. Res. Develop.* **4**(1), 16 (1965).
22. S. Sterman and J. G. Marsden. SPI, 18th Ann. Tech. Conf. Reinf. Plast. 1-D (1963).
23. J. E. Moreland. Tech. Bulletin, Malvern Min. Prod., Hot Springs, Ark. (1968).
24. M. W. Ranney, S. E. Berger, and J. G. Marsden. In *Composite Materials*, E. P. Plueddemann, Ed., Vol. 6, *Interfaces in Polymer Matrix Composites*, Academic Press, New York, Chapter 5 (1974).
25. E. P. Plueddemann. SPI, 20th Ann. Tech. Conf. Reinf. Plast. 19-A (1965).
26. Y. L. Fan and R. G. Shaw. SPI, 25th Ann. Tech. Conf. Reinf. Plast. 16-A (1970).
27. G. A. McFarren, T. F. Sanderson, and F. G. Schappell. SPE Ann. Tech. Conf., Atlantic City **34**, 19 (1976).
28. E. P. Plueddemann and G. L. Stark. SPI, 35th Ann. Tech. Conf. Reinf. Plast. 20-B (1980).
29. W. T. Collins and J. L. Kludt. SPI, 30th Ann. Tech. Conf. Reinf. Plast. 7-D (1975).
30. B. Arkles. 69th Ann. Conf. Am. Inst. Chem. Eng. (1976).
31. B. Arkles, J. Steinmetz and J. Hogan. 42nd Annu. Conf. Composites Inst., SPI, 21-C (1987).
32. O. J. Onufer and E. C. Staley. *Plast. Des. Process.* **7**, 9 (1976).
33. R. Longworth. In *Ionic Polymers.* L. Holliday, Ed., pp. 69–172, Wiley, New York (1975).
34. E. P. Plueddemann. *J. Adhesion Sci. Technol.* **3**(2), 131 (1989).
35. E. p. Plueddemann and W. T. Collins. In *Adhesion Science and Technology*, L. H. Lee, Ed., Vol. 9A, p. 329, Plenum, New York, 1975.
36. T. G. Waddell, D. E. Leyden, and D. M. Hercules. In *Silylated Surfaces*, D. E. Leyden and W. Collins, eds., Gordon and Breach Science Publishers, New York (1980).

# 8 | Other Applications of Silane Coupling Agents

## 8.1. General

In addition to their applications as coupling agents in mineral-filled organic resin composites, the various organofunctional silanes are arousing much interest in other areas of surface modification. A symposium on silylated surfaces, held at the Midland Macromolecular Institute in 1978, was devoted to some of these "other" applications of silylated surfaces.[1] Many additional properties of silylated surfaces present obvious possibilities, but have not been studied enough for practical application.

Surfaces of oxide minerals may be modified with monolayers of organofunctional silanes to make them hydrophobic or hydrophilic; and neutral, cationic, or anionic. Such control of mineral surfaces should be useful in oil recovery, ore flotation, pigment dispersion, and surface modification of metals.

Polyvalent cations or anions may be bound to surfaces through chelate-functional silanes with a mobility controlled by pH. This capability should be useful in medicine for controlled release or removal of ions from the body, in corrosion resistant coatings, and in commercial removal or recovery of ions from dilute solution.

Cellulose in many ways resembles silica with a hydroxylated surface having an isoelectric point of 2. Many of the applications of silanes on silica should be adaptable to paper, cotton, and wood.

Organofunctional silanes have an advantage over comparable organic compounds in that the silanes have potential for bonding to hydroxylated surfaces through several mechanisms. Electrokinetic forces can be used to attract or repel at relatively great distances. At closer approach, Van der

Waal's forces, hydrogen bonding, and covalent bonding are possible through silanol groups. By combining all of these possibilities in a single molecule, the silanes are uniquely capable of competing with water for hydroxylated surfaces.

## 8.2. Alkyl-Silylated Surfaces

Silanes can alter the critical surface tension ($\gamma_{crit}$) of a substrate as shown by contact angle measurements. Problems associated with such measurements are mentioned in 4.5.3, but $\gamma_{crit}$ of a surface is a convenient guide for predicting wettability or release qualities of a substrate. A liquid with a given surface tension will wet any surface with a higher critical surface tension. Critical surface tensions of typical construction materials are compared with silylated surfaces in Table 8.1.

### 8.2.1. Hydrophobing Agents

Alkylsilylating agents convert surfaces to water-repellent, low-energy surfaces useful in water-resistant treatments for masonry, electrical insulators, packings for chromatography, and in noncaking fire extinguishers.

Methylchlorosilanes react with water or hydroxyl groups at the surface to liberate HCl and deposit a very thin film of methylpolysiloxanes which has a very low critical surface tension and is therefore not wetted by water. Ceramic insulators may be treated with methylchlorosilane vapors or solutions in inert solvents to maintain high electrical resistivity under humid conditions.[2] The corrosive action of the evolved HCl can be avoided by prehydrolyzing the chlorosilanes in an organic solvent and applying them as organic solutions of organopolysiloxanols.

Hydrolyzed methylchlorosilanes are also dissolved in aqueous alkali and applied as aqueous solutions of sodium methylsiliconates. The siliconates are neutralized by carbon dioxide in the air to form an insoluble, water-resistant methylpolysiloxane film within 24 hr. Treatment of brick, mortar, sandstone, concrete, and other masonry protects the surface from spalling, cracking, efflorescence, and other types of damage caused by water.

Very water repellent films were obtained by depositing hexamethyldisiloxane on silicon chips in a high-energy plasma at low temperature.[3] Drops of water had contact angles as high as 180°. Furthermore these drops rolled off slightly inclined surfaces, suggesting a near-zero work of adhesion across the interface.

**Table 8.1.** Critical Surface Tensions ($\gamma_{crit}$)
of Selected Surfaces

| Surface | $\gamma_{crit}$ |
| --- | --- |
| Polytetrafluoroethylene | 18.5 |
| Methyltrimethoxysilane | 22.5 |
| Vinyltriethoxysilane | 25.0 |
| Paraffin wax | 25.5 |
| Ethyltrimethoxysilane | 27.0 |
| Propyltrimethoxysilane | 28.5 |
| Glass, soda-lime (wet) | 30.0 |
| Polychlorotrifluoroethylene | 31.0 |
| Polypropylene | 31.0 |
| Polyethylene | 33.0 |
| Trifluoropropyltrimethoxysilane | 33.5 |
| 3-(2 aminoethyl)-aminopropyltrimethoxysilane | 33.5 |
| Polystyrene | 34.0 |
| Cyanoethyltrimethoxysilane | 34.0 |
| Aminopropyltriethoxysilane | 35.0 |
| Polyvinylchloride | 39.0 |
| Phenyltrimethoxysilane | 40.0 |
| Chloropropyltrimethoxysilane | 40.5 |
| Mercaptopropyltrimethoxysilane | 41.0 |
| Glycidoxypropyltrimethoxysilane | 42.5 |
| Polyethyleneterphthalate | 43.0 |
| Copper (dry) | 44.0 |
| Aluminum (dry) | 45.0 |
| Iron (dry) | 46.0 |
| Nylon 6/6 | 46.0 |
| Glass, soda-lime (dry) | 47.0 |
| Silica, fused | 78.0 |

NOTE: Critical surface tensions for silanes refer to treated surfaces.

## 8.2.2. Chromatography

Silica packings for gas–liquid chromatography (GLC) columns are often treated with trimethylchlorosilane or other volatile silylating agents to reduce tailing of polar organic compounds. A chemically bonded methyl-silicone support allowed underivatized phenols to be analyzed without tailing.[4] This support was stable for temperature programming to 390°C and allowed elution of high-boiling molecules (e.g., hydrocarbons to the $C_{50}$ range).

Conventional thin-layer-chromatography (TLC) plates use a polar stationary phase (generally silica) and nonpolar solvent systems. Reversed-phase thin-layer chromatography (RPTLC) uses a nonpolar stationary phase and a polar solvent system (e.g., MeOH/H$_2$O). A favorite stationary phase

is silica that has been silylated with octadecyl trichlorosilane or other organic
trichlorosilanes. The principal separation mechanism of reversed-phase TLC
is the dispersion forces between the hydrocarbons of the sample molecules
and the alkyl groups on the treated silica. These forces depend in part on
the carbon chain length. For example, they are approximately twice as great
for octadecyl ($C_{18}$) groups as they are for octyl ($C_8$) groups.

Similar silylated silica may also be used as packing in gas chromatography and reversed-phase high-performance liquid chromatography (HPLC)
columns. Reversed-phase chromatography has become the most popular
branch of HPLC and owes its success to the development of stable bonded
hydrocarbon phases. Abel et al.[5] first used a silica support treated with
hexadecyltrichlorosilane as a bonded phase. Octadecyltrichlorosilane is the
favorite silylating agent at present, although trichlorosilane with ethyl,
phenyl, cyclohexyl, or octyl groups may give improved selectivity in some
applications.

When silica gel is treated with an alkyltrichlorosilane, the structure of
the bonded surface is poorly defined. It can be represented as an oriented
monolayer as shown in Figure 8.1, or by a thicker poly(alkylsiloxane) of
undefined structure. Obviously, the preparation of the silica gel, the amount
of water present, the nature of the coupling agent, and the conditions of
the reaction have a great influence on the morphology of the bonded phase.

Isolated silanol groups remaining on silica or in the coupling agent
layer may be silanated by a second treatment with small reactive molecules
like trimethylchlorosilane or hexamethyldisilazane to cover residual polar
sites on the solid phase.

Even though the silanating agent is bonded to silica through thermally
stable siloxane bonds, these bonds are hydrolyzable, and the rate of hydrolysis increases dramatically with increasing temperature and pH. Chromatographic operations at high pressure and elevated temperature at pH above
7 result in rapid deterioration of columns in contact with aqueous eluents.

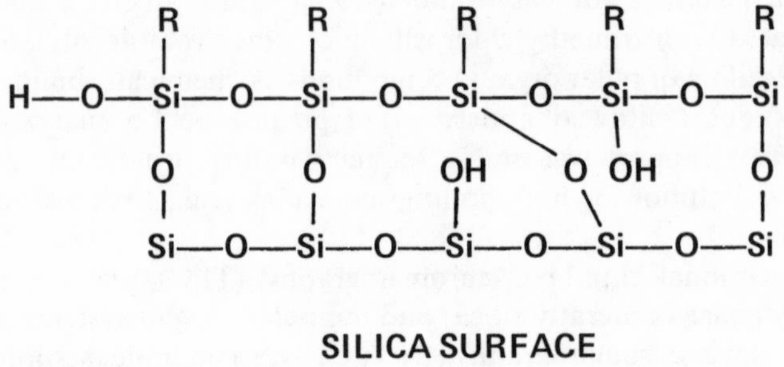

**SILICA SURFACE**

**Figure 8.1.**  Idealized monolayer of condensed coupling agent on silica.

The stability of the bound phase could probably be increased by additional cross-linking of the siloxane layer through C—C bonds as proposed by Legally.[6] Terpolymers of vinylstearate, vinyltriethoxysilane, and maleic anhydride were soluble in hot water (neutralized to pH 6) and provided stable lubricating films on glass bottles after a cure in steam.

Silica gel may also be treated with glycidoxypropyltrimethoxysilane followed by hydrolysis of the epoxy groups to obtain diol surfaces useful for protein liquid chromatography. Such columns frequently have a much higher chromatography capacity factors for nonpolar compounds than expected. Bogart et al.[7] studied the hydrolysis of epoxy groups on treated silica by solid-state NMR with cross-polarization and magic angle spinning. The results suggest that hydrolysis of epoxy groups is much more difficult than expected, and that as little as 10% unhydrolyzed epoxy on the surface significantly affects the capacity factor.

Pidgeon and Venkataram[8] bonded lecithin analogs to silica supports to mimic cell membranes in separation of many biomolecules. Lecithin was bonded to silica in a four-step process. Dodecane dicarboxylic anhydride was synthesized from 1,12-dodecanedicarboxylic acid using dicyclohexyl-carbodiimide as coupling agent. Monomyristoyl lysolecithin was reacted with the anhydride to form a lecithin carboxylic acid. That compound was reacted with carbonyldiimidazole to form lecithin-imidazolide. Finally, the imidazolide was coupled to the amino groups of commercial aminopropyl-silane-treated silica (Nucleosil-300, $7NH_2$). They found that chromatography columns packed with this lipid-bonded silica separated cytochrome P-450 in a single pass in a form 100 times more active than the same protein separated by other means (Figure 8.2).

### 8.2.3. Soil Treatment for Water Harvesting

Many drier areas of the world could be utilized for grazing if water were available for the livestock. Most of the rainfall received in these areas soaks into the upper soil layer, only to be quickly lost by evaporation. Water harvesting techniques which prevent this initial shallow infiltration now show promise of permitting increased development of many of these relatively dry areas. (Water harvesting is the collection of rainwater from land areas specifically treated to increase precipitation runoff.)

Myers[9] reviewed extensive studies by the U.S. Department of Agriculture on treatment of Arizona soil for water harvesting. Various latex polymers looked promising for soil consolidation, but imparted little water repellency. Aliphatic quaternary ammonium salts rendered soil water repellent, but generally did not penetrate, and erosion soon removed the treatment. An aqueous solution of sodium methylsiliconate penetrated the soil well and

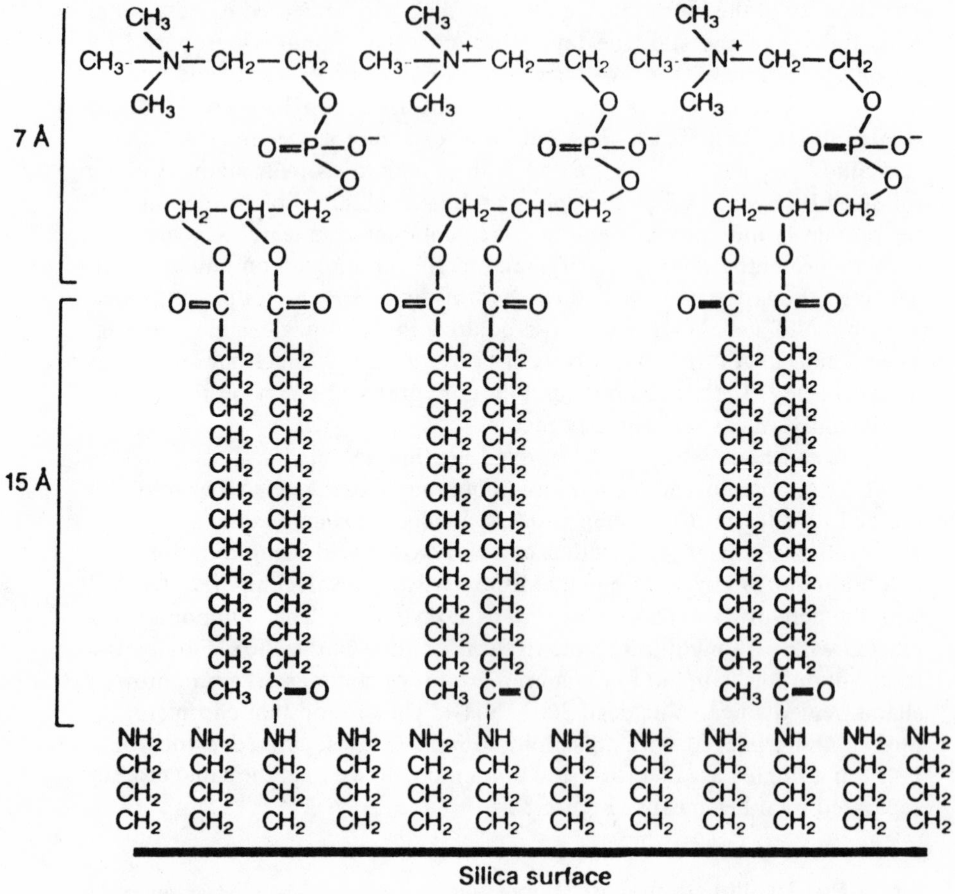

**Figure 8.2.** Cell membrane mimic is bonded to silica through aminopropyl silanes. Courtesy American Chemical Society.

provided water repellency during several years of outdoor exposure but the treatment lacked soil-binding properties and suffered from erosion.

Silicones as hydrophobing agents have outstanding weather resistance due to the polymer backbone of alternating silicon and oxygen atoms. Alkyl silicones provide low critical surface tension with complete freedom from imparted color, odor, or taste. Since low concentrations of silicones have no soil-binding properties, they should be mixed with appropriate organic polymeric materials as soil binders.

Sandy soil was treated effectively with an aqueous spray containing a polymer latex and a silicone water repellent.[10] The polymer latex served to bind the soil, while the water repellent caused run-off of the water. Among commercial polymers tested, an acrylic and a styrene–butadiene resin (SBR) latex looked most promising. Both a sodium methylsiliconate and a silicone oil emulsion were effective as water repellents, but the emulsion gave somewhat better results and was much more convenient to use. A mixture of SBR latex and silicone emulsion is stable at 50% solids, and may be diluted with water to about 3% solids for application to soil. The silicone emulsion, by itself, did not impart hydrophobicity to treated soil, but apparently formed an interpenetrating polymer network with the organic phase as the emulsion dried on the treated surface.

A mulch of natural soil formed into hydrophobic clods was proposed by Hillel[11] as a means of conserving irrigation water or natural soil moisture in areas of low precipitation. Clods of soil treated with a silicone-modified polymer latex were employed to form the top layer of soil. The treated clods allowed maximum intake of water, but constituted a barrier that reduced water evaporation.

Nitrogen and potassium leaching from fertilized sandy soils was reduced by a mulch of water repellent soil.[12] Sweet corn yields obtained with a sodium methylsiliconate spray on sandy soil were comparable to those obtained by protecting nitrogen and potassium with a plastic mulch.

Reclamation of land covered with mine spoil wastes was aided by treating a ring of soil around planted shrubs and trees with a silicone-modified latex[13] or ground paraffin wax. Good consolidation of soil and hydrophobicity was obtained by applying $3 \, l/m^2$ of a 6% silicone-latex emulsion.

### 8.2.4. Detergents

One of the major effects of silylating a solid particulate surface is to change the isoelectric point of the surface (IEPS). Washing of fabrics involves a complex interplay of IEPS of fabrics as modified by hardness ions, and charges on emulsified soil particles. A simple detergent formulation is proposed that comprises anionic or nonionic surfactants and fine insoluble particulates made strongly cationic by suitable silylating agents. Soil emulsified by the surfactant has a low IEPS and is picked up by the silylated particulates rather than by the fabric and is rinsed from the wash.[14]

A favorite particulate was diatomaceous earth (Filtercel®) treated with a hydrophilic quaternary ammonium functional silane,

$$(CH_3O)_3SiCH_2CH_2CH_2\overset{+}{N}-Cl^-CH_2CH_2OH$$
$$\underset{(CH_3)_2}{|}$$

**Table 8.2.** Comparison with Commercially Available Detergents

| | Final pH of wash water | Percent soil removed | | | Redeposition Index |
| | | Wool | Cotton | 50/50 cotton polyester | Polyester |
|---|---|---|---|---|---|
| Detergent | | | | | |
| Amway® [a] | 11.2 | 40 | 30 | 29 | 100 |
| Tide® [b] | 10.3 | 49 | 25 | 32 | 99 |
| Dash® [b] | 10.4 | 51 | 22 | 23 | 97 |
| Cold Water Surf® [c] | 9.4 | 57 | 24 | 28 | 97 |
| Dreft® [b] | 9.6 | 55 | 25 | 26 | 97 |
| Composition G | 8.2 | 44 | 34 | 35 | 100 |

[a] Amway Corporation, Ada, MI 49355.
[b] Procter & Gamble Company, Cincinnati OH 45201.
[c] Lever Brothers Company, New York NY 10022.

Standard soiled fabrics were washed in a Terg-O-Tometer in 1 liter of water containing 200 ppm hardness as $2:1$ $Ca^{2+}/Mg^{2+}$. The wash cycle consisted of 15 min agitation at 150 cycles/min at 120°F with two rinses. For commercial detergents, 0.8 g of product was used in each test. Detergents of this invention were composed of 0.15 g of Makon-10, 0.01 to 0.02 g of sodium silicate ($SiO_2/Na_2O$ ration 3.22) and 0.6 g of one of several types of insoluble particles treated with 1% by weight of $(CH_3O_3)_3Si(CH_2)_3N +$ $(CH_3)_2CH_2CH_2OH.Cl^-$. Composition G contained treated diatomaceous earth with an average particle size of 20 $\mu m$. The results are shown in Table 8.2.

## 8.3. Charge Transfer Chromatography

Separation of fused-ring or polynuclear aromatic hydrocarbons is of importance in isolating individual carcinogens. Their separation is complicated by the lack of any suitable functional groups other than the aromatic moiety. Polynuclear aromatic hydrocarbons can enter into charge-transfer interactions with polynitroaromatic electron acceptors. A bound phase with charge-transfer acceptor groups was prepared by treating aminopropyl-silylated silica with 3-(2,4,5,7-tetranitrofluorenone).

Liquid chromatography of polynuclear aromatic compounds on this bonded phase gave good selectivity, but rather poor efficiency.[15] Efficiency of separation was improved by decreasing the functionality of the bound phase. Silica was silylated with a mixture of 80% alkylsilane and 20% aminopropylsilane and the amine groups converted to the 3-(2,3,4,7-

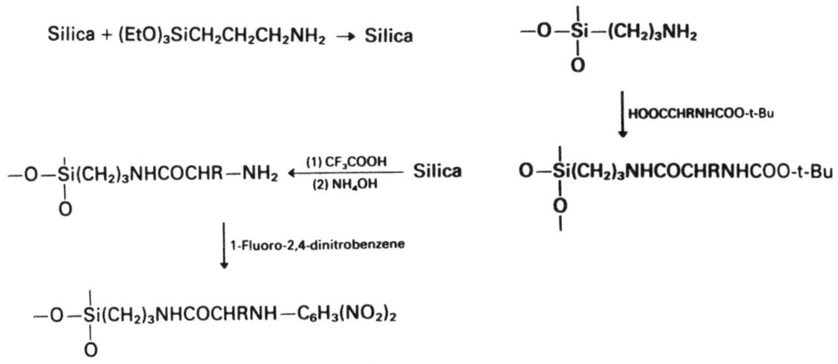

Silica + (EtO)$_3$SiCH$_2$CH$_2$CH$_2$NH$_2$ → Silica $\quad$ —O—Si—(CH$_2$)$_3$NH$_2$
$\qquad\qquad\qquad\qquad\qquad\qquad\qquad\qquad\qquad\qquad$ |
$\qquad\qquad\qquad\qquad\qquad\qquad\qquad\qquad\qquad\qquad$ O

$\qquad\qquad\qquad\qquad\qquad\qquad\qquad\qquad\qquad$ ↓ HOOCCHRNHCOO-t-Bu

$\qquad\qquad\qquad\qquad\qquad\quad$ (1) CF$_3$COOH
—O—Si(CH$_2$)$_3$NHCOCHR—NH$_2$ ← $\overline{\qquad\qquad\qquad}$ Silica $\quad$ O—Si(CH$_2$)$_3$NHCOCHRNHCOO-t-Bu
$\quad$ |$\qquad\qquad\qquad\qquad\qquad$ (2) NH$_4$OH $\qquad\qquad\qquad\qquad$ |
$\quad$ O $\qquad\qquad\qquad\qquad\qquad\qquad\qquad\qquad\qquad\qquad\qquad\quad$ O

$\quad$ ↓ 1-Fluoro-2,4-dinitrobenzene

—O—Si(CH$_2$)$_3$NHCOCHRNH—C$_6$H$_3$(NO$_2$)$_2$
$\quad$ |
$\quad$ O

| Amino acid | R-groups on α-carbon |
|---|---|
| glycine | —H |
| L-alanine | —CH$_3$ |
| L-valine | —CH(CH$_3$)$_2$ |
| L-phenylalanine | —CH$_2$C$_6$H$_5$ |

**Figure 8.3.** Preparation of 2,4-dinitrophenyl(DNP)-modified solid phase.

tetranitrofluorenimine) derivative for separation of various substituted anthracenes and chrysenes.[16]

Another way to control the Π-complexing strength of the bonded phase was to apply 1-fluoro-2,4-dinitrobenzene to amino acid-bonded phases on silica as outlined in Figure 8.3.

Retention of 1-aza[6]helicene racemates on the dinitrophenyl-modified amino acid-bonded phases are shown in Table 8.3. Only the L-alanine-bonded phase gave any separation of optical isomers. The same series of bonded phases were compared with aminofunctional phases in separation

**Table 8.3.** Data for the 1-AZA[6]Helicene Racemate on DNP(2,4-Dinitrophenyl)-Amino Acid Bonded Phases[a]

| Bonded phase | Tr (min) | $k'$ | $k'$/meq/g | $\alpha$[b] |
|---|---|---|---|---|
| DNP–glycine | 12.44 | 15.70 | 33.4 | — |
| DNP–L-alanine | 10.20 | 11.75 | 29.38 | 1.06 |
| | 10.74 | 12.42 | 31.05 | — |
| DNP–L-valine | 8.10 | 7.53 | 14.76 | — |
| DNP–L-phenylalanine | 10.97 | 12.38 | 24.76 | — |

[a] Mobile phase: 1.5% (v/v) acetonitrile in isooctane; temperature: 26.5°C.
[b] $\alpha$ is the selectivity obtained from $k'_2/k'_1$, where $k'_2$ is the most-retained species and $k'_1$ is the least-retained.

**Table 8.4.** Capacity Factors for Polynuclear Hydrocarbons
(cyclohexane mobile phase at 20°C)

| Solute | $k'$ values on bonded phases on silica[a] | | | | | | |
|---|---|---|---|---|---|---|---|
|  | 1 | 2 | 3 | 4 | 5 | 6 | 7 |
| Naphthalene | 0.68 | 0.25 | 0.84 | 0.76 | 0.80 | 0.68 | 0.47 |
| Fluorene | — | 0.58 | — | 1.82 | 1.95 | 1.67 | — |
| Anthracene | 2.27 | 0.84 | 5.93 | 4.72 | 3.68 | 3.45 | 2.10 |
| Phenanthrene | 2.56 | 0.93 | 6.04 | 4.92 | 4.06 | 3.98 | 2.22 |
| Pyrene | 3.36 | 1.28 | 13.57 | 9.73 | 8.40 | 8.44 | 4.48 |

[a] 1 = amine (aminopropylsilane)
2 = L-phenylalanine-amine
3 = DNP-amine
4 = DNP-glycine-amine
5 = DNP-L-alanine-amine
6 = DNP-L-phenylalanine-amine
7 = DVP-L-valine-amine

of various polycyclic aromatics in cyclohexane as the mobile phase
(Table 8.4).

## 8.4.  Bound Chelate-Functional Silanes

### 8.4.1.  General

Organofunctional silanes with chelating groups are available commercially or may be prepared from commercial silanes. Chelate-functional silanes may be prepared separately and applied to a siliceous substrate from organic solvent, or a silane-treated surface may be modified chemically to prepare immobilized chelating agents.[17] Such immobilized chelating agents may be used to preconcentrate ions from dilute solutions for analytical purposes, or for large-scale removal and recovery of ions from solution. Metal ions bonded to immobilized chelating agents may be used as the bound phase in high-performance liquid chromatography or as heterogeneous catalysts.

Chelate-functional groups bonded to inorganic supports have several advantages over comparable functional organic resins. The inorganic supports have greater mechanical strength which allows higher operating pressure and flow rates. Moreover, the functional groups are concentrated at the surface, which allows higher reaction rates than can be obtained in a process involving diffusion through a resin. Many inorganic supports are much less costly than organic resins. The need has been expressed repeatedly for more stable bonding through organofunctional silanes to minerals, especially at pH above 7.

### 8.4.2. Ligand-Exchange Chromatography

Chromatographic separation of aminosubstituted organic molecules is enhanced by metals ions, ether on the support or in the mobile phase, by a process termed ligand exchange chromatography (LEC). Silica gel impregnated with metal salts has been used as the stationary phase, but this system lacked stability and reproducibility.[18] Metal ions may also be bonded to chelate-functional silanes on a porous silica support. This approach retains the high selectivity of LEC on the one hand, and enhances the efficiency of separation on the other.

A $Co(en)_2^{+3}$ bonded phase was prepared by silylating silica gel with an ethylenediamine-organofunctional silane and contacting it was $Co(en)_2Cl_2^+$. This bonded phase was highly stable on silica and formed complexes with nucleotides. Mg(II) was added to the mobile phase to form innersphere complexes with the nucleotides to compete with bonded $Co(en)_3^{+3}$. Using such a system, an isocratic separation of ten nucleotides and nucleosides was accomplished in less than 18 min.[19]

Other chelate-functional stationary phases were prepared by reacting aminopropyl-silylated silica was $CS_2$ (dithiocarbamate) or with ethyl benzoylacetate (diketone). Cu(II) was used to treat the columns for separation of 22 aromatic compounds[20] as shown in Table 8.5.

### 8.4.3. Preconcentration of Aqueous Ions

Silica gel that is silylated with silane reagents which contain, or can be modified to contain, a functional group capable of chelation with metal ions can be of analytical use. Leyden[21] treated silica gel and controlled pore glass with $N$-$\beta$-aminoethyl-$\gamma$-aminopropyltrimethoxysilane in toluene for preconcentration of anions and cations.

A general procedure was to wash the substrate with 2 $M$ nitric acid, rinse with distilled water, and dry in an oven. The dried substrate (5 g) was stirred with 10 ml of a 10% solution of silane in toluene at room temperature. The treated substrate was recovered by filtration and rinsed with toluene and isopropyl alcohol. The product was dried in an 80°C oven for 8–12 hr and stored in a desiccator until used.

The pH dependence of ion take-up was determined by stirring 200 mg of treated solid with 2 $\mu$mol of a 0.1 $M$ solution of the ion of interest at pH adjusted with 0.1 $M$ HClO$_4$ or 0.1 $M$ NaOH. The relative up take was measured directly on the substrate by X-ray fluorescence,[22] or the ion could be eluted by an appropriate solvent (1 $M$ HCl for copper II, and 4 $M$

**Table 8.5.** $k'$ Values on Diketone-Bonded Column
[mobile phase: hexane/$CH_2Cl_2$/MeOH (7:1:0.5)]

| Compound | No Cu(II) | With Cu(II) |
|---|---|---|
| 1. Naphthalene | 0.89 | 0.56 |
| 2. 1-Aminonaphthalene | 0.89 | 4.16 |
| 3. 2-Aminonaphthalene | 0.89 | 9.58 |
| 4. 1-Aminoanthracene | 1.11 | 5.08 |
| 5. 2-Aminoanthracene | 1.32 | 9.32 |
| 6. 2-Amino-9-fluorenone | 1.60 | 7.46 |
| 7. 4-Amino-9-fluorenone | 1.72 | 4.34 |
| 8. 2-Amino-9-hydroxyfluorene | 2.42 | 38.7 |
| 9. 9-Fluorenone | 0.36 | 0.56 |
| 10. Fluorene | 0.14 | 0.22 |
| 11. 2-Aminofluorene | 0.92 | 13.6 |
| 12. o-Chloroaniline | 0.58 | 1.40 |
| 13. m-Chloroaniline | 1.00 | 4.76 |
| 14. p-Chloroaniline | 1.00 | 7.88 |
| 15. o-Nitroaniline | 1.53 | 2.84 |
| 16. m-Nitroaniline | 2.28 | 6.68 |
| 17. p-Nitroaniline | 9.88 | 10.7 |
| 18. o-Toluidine | 0.53 | 3.12 |
| 19. m-Toluidine | 0.53 | 7.56 |
| 20. p-Toluidine | 0.53 | 12.36 |
| 21. o-Anisidine | 0.50 | 4.44 |
| 22. p-Anisidine | 0.84 | 33.6 |

NH$_4$OH for molybdate), and the ion determined in the solution by atomic absorption spectroscopy.

Equilibration of metal ions with the immobilized diamine group is reached in a matter of minutes rather than hours, as required by polymer-based chelating ion exchangers. The percent metal ion extracted as a function of pH for several metal ions on diamine-silylated silica gel is shown in Figure 8.4.

An advantage of the diamine-functional group which was not expected is the ability of the protonated form to interact with anions.[23] Certain anions form strong ion-pairs or salts with the protonated diamine group, causing them to be extracted from solution onto the surface. Especially strongly bound are the oxyanions of molybdenum, tungsten, selenium, and arsenic. The pH dependence of the extraction of molybdate is shown in Figure 8.5. In strongly acidic solutions, the formation of molybdic acid weakens the ion pair, and in basic solutions there is no charge on the diamine to retain the anion.

Removal of ions on silylated silica could become a commercial process for reducing residual ionic impurities, or for commercial recovery of metals

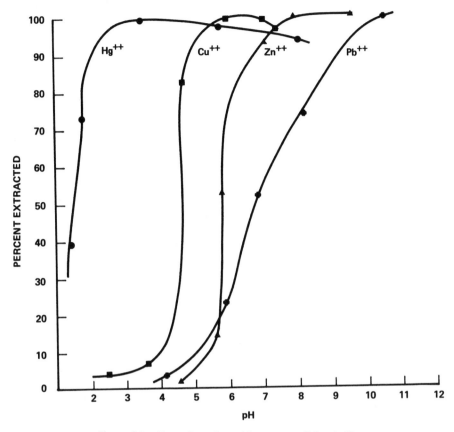

**Figure 8.4.** Extraction of metal ions on en silylated silica.

from dilute solution (e.g., uranium for sea water). Such production will require high flow rates and capability for recycling. Treated silica provides high flow rates because functional groups are on the surface of the matrix. Ion-exchange resins are orders of magnitude slower because diffusion of ions is so much slower through a cross-linked resin. Newer macroreticular resins contain large, deep pores in spherical beads with functional groups on the surface of these pores. Silica gel still has a great advantage in available surface area and cost, if the silylated surface can be made permanent to recycling.

The permanence of chelate-functional silanes on a silica surface was best tested in a continuous apparatus that could provide a large number of elutions automatically.[24] Various aminofunctional silanes were compared with an 8-hydroxyquinoline-functional silane in removal of Cu(II) with silylated silica gel.[25]

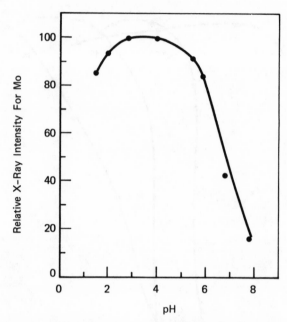

**Figure 8.5.** Extraction of molybdate ion on en silylated silica.[19] Courtesy of *Analytical Chemistry.*

A simple procedure of preparing monomeric trimethoxysilylpropyl-substituted phenols[26] was adapted to the synthesis of 7-trimethoxy-silylpropyl-8-hydroxyquinoline by the following series of reactions:

1. 8-Hydroxyquinoline + allylbromide (KOH) → allyl ether(1).
2. (1) heated, rearranged (Claisen) to a 7-allyl-8-hydroxyquinoline.
3. The phenol group was blocked by silylating with hexamethyl-disilazane.
4. Trimethoxysilane was added to the allyl group in the presence of a Pt catalyst.
5. Silica gel was treated with a 2% solution in toluene of the crude reaction product.

### 8.4.4. Continuous Measurements

For continuous operation, an Altex-Beckman high-performance liquid chromatograph, complete with a 110-A pump, a six-way solvent switching valve, a Hitachi model 100-10 variable wavelength spectrophotometer, a

**Table 8.6.** Sequence of Solutions Flowing Through the
Column in Cu(II) Capacity Determinations

| Order | Solution | Volume (mL) |
|-------|----------|-------------|
| 1 | Water | 5 |
| 2 | 0.2 M NaOAc buffer | 5 |
| 3 | 0.01 M Cu(II) solution pH 5 | 10 |
| 4 | Water | 5 |
| 5 | Dilute nitric acid pH 2 | 5 |
| 6 | Water | 5 |

Kip and Zonnen BD 40 recorder, and a microprocessor, was utilized. High-pressure stainless steel columns, 250 mm long and 2.1 mm inner diameter (i.d.), were weighed, dry-packed with approximately 0.2 g of sample, and then weighed again to obtain the exact weight of the packing. A solution of Cu(II) flowed from the pump and through the column. The Cu(II) was bound by the immobilized silane until the apparent Cu(II) capacity of the substrate was exceeded.

A post column reactor system was coupled to the liquid chromatograph. A colorimetric reagent, (1,2-cyclohexylenedinitrilo)-tetraacetic acid

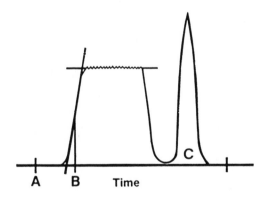

A. Start of copper (II) flow

B. Half Maximum of copper (II) - DCYTA
   complex breakthrough peak.

C. Acid eluted copper (II) - DCYTA complex peak.

**Figure 8.6.** Typical data obtained during capacity studies Courtesy Gordon and Breach Science Publishers.

**Table 8.7.** Summary of Stability Data Continuous Cycling Organofunctional Silanes on Silica (silanes applied from 4% solution in toluene)

| Proposed structure on silica | Abbreviation for silane | Initial capacity mmol $Cu(II)$ $g^{-1}$ | Fraction remaining | |
|---|---|---|---|---|
| | | | after 8 cycles | after 13 cycles |
| Si—O—Si(CH₂)₃NH₂ <br> Si—O—Si(CH₂)₃NH₂ | APT | 0.045 | 0.73 | 0.58 |
| Si—O—Si(CH₂)₃NHCH₂CH₂NH₂ <br> Si—O—(CH₂)₃NHCH₂CH₂NH₂ | AEAPT | 0.266 | 0.75 | 0.69 |
| Si—O—Si—(CH₂)₃—NHCH₂CH₂NH <br> CH₂ <br> CH₂ <br> Si—O—Si—(CH₂)₃—NHCH₂CH₂NH | bis-TETA | 0.074 | 1.00 | 1.00 |
| Si—O—Si(CH₂)₃— <br> Si—O—Si(CH₂)₃— | HQPT | 0.019 | 0.99 | 0.84 |
| Untreated silica | — | 0.004 | — | — |

$^a$ AEAPT applied from 1% solution in toluene.

(DCYTA), flowed from a Ranin peristaltic pump to a Teflon three-way mixing chamber located below the column. After the eluent was mixed with the colorimetric reagent, the absorbance at 715 nm was measured by passing the mixed solution through a 20 µl flow cell in a spectrophotometer. Back pressure in the system was maintained by utilizing a 1 m length of 0.33-nm-i.d. Teflon tubing connected to the exit of the flow cell. the sequence of solutions flowing through the column is shown in Table 8.6. Total cycle time was 70 min, and could be set to repeat indefinitely without supervision. Figure 8.6 shows an example of data obtained in a typical cycle.

Silica gel used in continuous cycling tests had a surface area of $290 \, m^2 g^{-1}$ and was treated with 4% (v/v) silane solution in toluene for 30 min at room temperature. One sample, for comparison, was treated with a 1% (v/v) solution of AEAPT in toluene.

Relative retention of Cu(II) capacity was shown by comparing capacity/initial capacity of the various silylated silicas as a function of cycle number (Table 8.7).

### 8.4.5. Immobilized Metal Complex Catalysts

Transition-metal complexes have been bonded chemically to the surface of insoluble macromolecules. Such complexes have been shown to produce catalyst systems that combine the advantages of homogeneous catalysis of high activity and selectivity with the ease of handling associated with heterogeneous catalysts. Most work in this field has concerned organic polymers as the macromolecular support, whereas inorganic macromecules, such as silica, have physical properties that offer considerable advantages as supports for large-scale applications of heterogeneous catalysts. Because organic polymers lack a rigid structure, their conformation and hence the shape and size of the particles is strongly influenced by solvents, temperature, and pressure. This may limit their use in large-scale apparatus where long catalyst beds and large pressure drops are encountered. Silica, on the other hand, is mechanically rigid and is unaffected by all but the most severe solvent and temperature conditions.

Two general methods (see below) are available for immobilizing metal complex catalysts on metal oxide surfaces by the use of ligand silane coupling reagents of the type $X_3SiL$ where X is a hydrolyzable group and L is a liganding group.[27]

In Method A, the ligand silane is used to react with surface hydroxyl groups to form a ligand-functional silica, and then a metal complex precursor is allowed to react with the funtionalized surface. In Method B, a metal–ligand silane complex is first formed in solution and then allowed to react with surface hydroxyls of the support.

## Silylating Agents
### Method A

$$\equiv SiOH \ + \ X_3SiL \ \rightarrow \ \equiv SiOSi-L + HX \qquad (1)$$

surface silanol    ligand silane         ligand silica

$$\equiv SiOSi-L + \qquad M \qquad \rightarrow \qquad \equiv SiOSiLM \qquad (2)$$

ligand silica    complex precursor    surface complex

### Method B

$$X_3SiL \ + \qquad M \qquad \rightarrow \qquad X_3SiLM \qquad (3)$$

ligand silane    metal precursor    metal–silane complex

$$\equiv SiOH \ + \qquad X_3SiLM \qquad \rightarrow \ \equiv SiOSiLM + HX \qquad (4)$$

surface silanol    metal–silane complex

The British Petroleum[27] group has pointed out advantages and disadvantages of both procedures.

Tertiary phosphine ligand silanes were prepared by the following different procedures:

1. $(EtO)_3SiCH{=}XH_2 + HPPh_2 \xrightarrow{\text{u.v.}} (EtO)_3SiCH_2CH_2PPh_2$
2. $Cl_3SiH + p\text{-}CH_2{=}CHCH_2CH_2C_6H_4PPh_2 \rightarrow$
   $p\text{-}Cl_3Si(CH_2)_4C_6H_4PPh_2$
3. $-OSiCH_2CH_2CH_2Cl + KPPh_2 \rightarrow$
   $-OSiCH_2CH_2CH_2PPh_2$   on silica

Silica treated with these ligand silanes may contain residual silanol groups. The accessible groups that remain may be removed by silylating with nonligand silanes such as trimethylchlorosilane or hexamethyldisilazane. Such removal has been shown to reduce side reactions of the complex surface catalysts at high temperaures.

Applications of supported metal complex catalysts were reviewed by Pinnavaia[28] (Table 8.8). Some newer work describes quaternary ammonium salt-functional silanes on silica gel as effective heterogeneous catalysts for continuous redistribution of trichlorosilane to dichlorosilane and silicon

**Table 8.8.** Representative Silica-Supported Metal Complex Catalysts Prepared by Use of Ligand Silane Coupling Agents[a]

| Surface complex catalyst[b] on silica | Method of preparation | Catalytic application |
|---|---|---|
| $CpTiCl_2(C_5H_4)-$ | B | Hydrogenation |
| $TiCl_2(C_5H_4)_2-$ | B | Hydrogenation |
| $RhCl[PPh_2(CH_2)_2]_3-$ | A, B | Hydrogenation and hydro-Formation |
| $RhCl[(CO)PPh_2(CH_2)_2]_2-$ | B | Hydroformylation |
| | A | Carbonylation |
| $Rh(acac)(CO)PPh_2(CH_2)_2-$ | B | Hydroformylation |
| $PhCl(PPh_2C_6H_4(CH_2)_4]_2-$ | B | Hydrogenation |
| $RhCl_3/PPh_2(CH_2)_2-$ | A | Hydrosilylation |
| $RhCl_3(COD)[PPh_2(CH_2)_{14}]-$ | A, B | Hydroformylation |
| $RhCl_3/NMe_2(CH_2)_3-$ | A | Hydrosilylation |
| $PtCl_4^{2-}/SnCl/PPh_2(CH_2)_2-$ | A | Isomerization |
| $H_2PtCl_6/NMe_2(CH_2)_3-$ | A | Hydrosilylation |
| $Pd(acac)_2/PPh_2(CH_2)_8-$ | A | Carbonylation |
| $Co(CO)_2(C_5H_4)-$ | A, B | Hydroformulation |
| $Co(acac)_2/PPh_2(CH_2)_2-$ | A | Hydroformylation |
| $NiCl_2[PPh_2(CH_2)_2]-$ | A | Oligomerization |
| $IrCl[PPh_2C_6H_4(CH_2)_4]-$ | A | Hydrogenation |

[a] Abbreviations used in this table are as follows: acac, acetyl-acetonate; COD, 1,5-cyclo-octadiene; Cp, cyclopentadienyl.
[b] In cases where undefined metal complexes were formed by Method A, the metal precursor/ligand silica system is specified.

tetrachloride. In a continuous reactor, the gas-phase reaction rate at 145°C was 25 to 70 times that of a comparable reactor containing a quaternary ammonium-functional ion-exchange resin.[29]

$$2HSiCl_3 \rightarrow H_2SiCl_2 + SiCl_4$$

## 8.5. Antimicrobials

Surface-bonded organosilicon quaternary ammonium chlorides have enhanced antimicrobial and algicidal activity.[30] Thus, the hydrolysis product of 3-(trimethoxysilyl)-propyldimethyloctadecylammonium chloride exhibited antimicrobial activity against a broad range of microorganisms while chemically bonded to a variety of surfaces. The chemical was not removed from surfaces by repeated washing with water, and its antimicrobial activity could be attributed not to a slow release of the chemical, but rather to the surface-bonded chemical.

White and Gettings[31] described several methods of testing antimicrobial properties of surfaces treated with the quaternary ammonium-functional silane. The $-silane-N^+-C_{18}H_{37}$ surface is hyrophobic and cationic. Bacteria are so small that they behave as colloidal particles and are attracted to surfaces by various means including hydrophobic attraction and electrostatic attraction. When bacteria contact a $-silane-N^+-C_{18}H_{37}$ surface they suffer rupture of cell membranes, leading to their destruction (Figure 8.7). In an aerosol test, swatches were inoculated with an aerosol of test bacteria at a level of about $1.3 \times 10^6$ organisms per swatch. Whatman No. 40 filter paper as a control was compared to ISO·BAC® nonwoven fabric

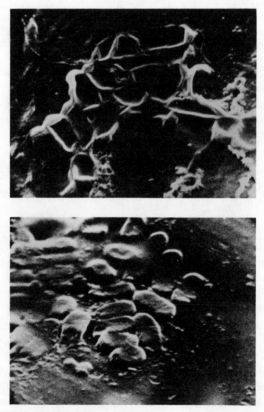

**Figure 8.7.**   Scanning electron micrographs of bacteria on fabrics.

**Table 8.9.** Aerosol Tests of Bacteria on Surfaces

| Aerosol treatment | Count/mL( × $10^{-3}$) at each dwell interval | | | |
|---|---|---|---|---|
| | 0 | 15 min | 1 hr | 3 hr |
| *Pseudomonas aeruginosa* | | | | |
| Control | 109 | 112 | 118 | 120 |
| ISO·BAC® | 15 | 0 | 0 | 0 |
| *Escherichia coli* | | | | |
| Control | 124 | 123 | 115 | 109 |
| ISO·BAC® | 18 | 0 | 0 | 0 |

treated with Dow Corning® 5700 antimicrobial agent (Table 8.9). Results are the average of triplicate tests.

The antimicrobial quaternary ammonium-functional silane may be used as the emulsifier to make stable emulsions of silicone fluids or mineral oil.[32] In these emulsions the methoxy groups of the silane are not hydrolyzed until a film of the emulsion is dried on a surface. Once the emulsion breaks on a surface, the emulsifier attaches to the surface through siloxane bond formation, and it is not longer available for rewetting or resolubilizing the deposited film.

## 8.6. Polypeptide Synthesis and Analysis

Silica or controlled pore glass supports treated with (chloromethyl)-phenylethyl silanes are replacing chloromethylated styrene divinylbenzene (Merrified Resin) as supports in polypeptide synthesis.

The silylated support is allowed to react with the triethylammonium salt of a protected amino acid. Once the initial amino acid residue has been coupled to the support, a variety of peptide synthesis methods can be used. Automation allows eight synthetic steps daily. At the completion of synthesis, the anchored peptide is separated from the support with hydrogen bromide in acetic acid.

Edman degradations can be accomplished by treating aminopropyl-silylated supports (a silica with pores of 10–75 Å yields about $2 \times 10^{-7}$ moles of aminopropylsilane per gram of glass) with the peptide to be analyzed in the presence of dicyclohexylcarbodiimide.[33] The carboxyl end of the peptide bonds to the amino group of the silane through an amide group. The bound peptide is then treated with phenylthiocyanate in the presence of base to yield a *N*-terminal phenylthiocarbamyl derivative which, on treatment with acid, cyclizes to a phenylthiohydantion and cleaves. The hydantion is analyzed and the process repeated with the bound peptide residue.

## 8.7. Immobilized Enzymes

### 8.7.1. General

For years the fermentation industry has depended on the catalytic properties of enzymes derived from microorganisms, but the microorganisms of fermentation usually produce a myriad of compounds from which the desired product must be purified. Because isolated enzymes are easier to handle, more specific in their function, and more predictable in their activity, they offer advantages over traditional fermentation processes.

Use of enzymes to catalyze reactions in cell-free systems has been limited by the difficulty of enzyme isolation, lability of the enzymes, and difficulty in effecting clean separations of enzymes from reaction mixtures. An approach that has circumvented some of these problems is to attach enzymes to solid support materials.[34,35]

### 8.7.2. Enzyme Supports

A suitable support material should be dimensionally stable, durable, hydrophilic (many enzymes denature at a hydrocarbon–water interface), regenerable, and have a high capacity for enzymes. Alumina, silica, glass, and other silicates have many advantages over organic polymeric supports if a suitable means of binding enzymes to minerals can be developed.

Enzymes adsorb directly on some minerals, but are relatively easily desorbed in water. Direct chemical modification of a silicate surface allows covalent bonding of enzymes to siliceous substrates. In water, the surface reverts rather easily to silicate structures. Silane coupling agents used as coupling agents in glass-reinforced plastics offer the possibility of more permanent bonds, and have been adapted as the general means of bonding enzymes to siliceous supports.

Numerous minerals have been used as substrates for enzymes. Iron oxide used as a support enables bound enzymes to be separated magnetically from reaction mixtures. Enzymes bonded to fine particulate minerals, like diatomaceous earth or aluminum oxide, may be used in a fluid bed reactor. Fluidization is achieved by passing a fluid-containing substrate upward through the bed of particulates fast enough to overcome gravitational forces on the particles, but slow enough to keep the particles from being transported out of the vessel.

Fine particulate solids cannot be used in packed bed reactors, since they reduce the rate of flow or generate excessive back pressure. Particle sizes in the range of 1 to 3 mm (25–75 mesh) appear to be most practical for packed bed reactors. For such particles to have a high capacity for enzymes, they must be highly porous. To allow entrance of large polymers like enzymes, a pore size in the range of 200 to 600 Å is desirable.[36] In extreme cases of very large enzyme complexes, a pore diameter greater than 1000 Å may be required. Controlled-pore glass (CPG), porous silica (PS), and certain controlled-pore ceramics developed by Corning Glass Works fall in this optimum range and they have become the preferred substrates for commercial-packed immobilized enzyme reactors.

### 8.7.3. Bonding Reactions

Amines are the favorite functional groups for bonding enzymes to glass. Chlorinated glass may be aminated directly[37]:

$$\equiv SiCl + 2NH_3 \rightarrow \ \equiv Si-NH_2 + NH_4Cl$$

or the glass may be treated with an aminopropyl silane coupling agent. Several reactions may be used to bond glass-amine ($G-NH_2$) groups to the amine groups ($E-NH_2$) or phenol groups $E-CH_2C_6H_3OH$ of enzymes[38]:

A. Michael coupling through gluteraldehyde:

$$G-NH_2 + HCO(CH_2)_3CHO + NH_2E \rightarrow$$

$$G-NH=CH(CH_2)_3-C=N-E$$

Stability is improved by reducing the imide carbon–nitrogen double bonds with sodium borohydride.

B. Amide coupling:

$$G-NH + \text{succinic anhydride} \rightarrow G-NHCOCH_2CH_2COOH$$

$$G-NHCOCH_2CH_2COOH + E-NH_2 \xrightarrow{\text{carbodiimide}}$$

$$G-NHCOCH_2CH_2CONH-E$$

C. Azo coupling:

$$G-NH_2 + nitrobenzoylchloride \rightarrow G-NHCOC_6H_4NO_2 \xrightarrow{[H]}$$

$$G-NHCOC_6H_4NH_2 \xrightarrow[HCl]{NaNO_2} G-NHCOC_6H_4N_2 + Cl^-$$

$$\xrightarrow{E-CH_2C_6H_4OH} G-NHCOC_6H_4-N=NC_6H_4CH_2E$$

For bonding of enzymes to glass, an aminofunctional silane has the advantage over direct amination of glass in that the ultimate bond to glass is siloxane rather than silazane. Siloxanes are generally more resistant to hydrolysis than are silazane bonds. Hydrolysis of siloxane bonds under acid or neutral conditions leaves silanol groups that remain in equilibrium with the siloxane structure. At a pH above 9, the siloxane bonds are attacked by alkali hydroxides to form soluble silicate or siliconate ions. This process can ultimately destroy not only the bond of enzyme to the silicate support, but the underlying silicate structure itself.

### 8.7.4. Stability of Immobilized Enzymes

The stability of aminopropylfunctional silane-treated controlled pore glass was studied by Royer and Liberatore[39] under continuous flow conditions over a pH range of 4 to 9. Amine remaining on the glass surface after pumping 1 liter of solution through a packed bed at a flow rate of 1 ml/min is shown in Figure 8.8. Retention of amine function on glass is poor at any pH above 6. Commercial experience has shown that enzymes that function at an acid pH have very respectable stability when bound to silanized glass. Bound glucoamylose, for example, has a half-life of 55 days during continuous operation at 45°C. Since porous glass substrate is the most expensive part of the bound enzyme system, it is recovered by ignition to destroy residual enzyme, followed by bonding a new enzyme to a regenerated carrier.

Several methods have been proposed to improve the stability of the bound organic phase. Saturation of alkaline solutions with sodium silicate has been observed to suppress the attack of alkali on glass.[38] Stability of the glass surface may also be improved by treating the glass with metal ions derived from chlorides of titanium, tin, zirconium, or vanadium.

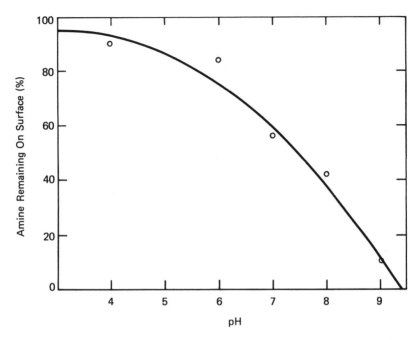

**Figure 8.8.** Stability of en silanized porous glass. The glass was washed in a column with 1 liter of buffer at a rate of 1 ml/min.

Stability of silane coupling agents on glass may be improved by additional cross-linking[40] (Section 5.2.4.3). Some siloxane cross-links are formed when organofunctional trialkoxysilanes are hydrolyzed and condensed on a mineral surface. It has been general practice to silanize porous glass with far more aminopropylsilane than is necessary for monolayer coverage. Excess coupling agent is cross-linked in the interphase region as a polysiloxane layer. Resistance to water is improved by baking this layer at elevated temperature, but organic reactivity is lost.

Royer and Libatore[39] attempted to improve the stability of enzyme supports by adsorbing organic polymers on aminopropyl-silanized glass and then cross-linking them by attaching the polymers at multiple points. Polyethyleneimine (PEI), polyacrylic acid (PAA), and poly[methylvinylether/maleic anhydride] (Gantrez®) were employed. The treated surfaces were used to immobilize the proteolytic enzyme trypsin. Several of the coated substrates had good stability, but only one had activity (units per milligram derivative) comparable to trypsin on commercial silanized porous glass. A unit of activity is 1 $\mu$mol of $N$-$\alpha$-benzoyl-L-arginene ethyl ester hydrolyzed per minute. The most favorable results were obtained with trypsin on porous silica that had been treated with PEI, cross-linked with gluteraldehyde, and had PAA subsequently added to it with a water-soluble carbodiimide.

Some of the concepts tried with immobilized metal ions on surfaces treated with chelate-functional silanes (Section 8.4.3) should be applicable to more stable bonding of enzymes.

### 8.7.5. Applications of Immobilized Enzymes

**8.7.5.1. Analytical.** Packed-bed immobilized enzyme reactors are used for routine chemical analysis in complex matrices. Enzymes can be extremely selective catalysts in terms of relative rates of closely related species, but many variables must be controlled to calibrate a given enzyme preparation. Because bound enzymes become a semipermanent part of the analyzer, they can be calibrated more easily than a soluble enzyme, and can then be used for multiple analyses. Highly selective and fast analyzers (e.g., 60 samples per hour), which require very small samples (typically 100 $\mu$l), have been developed.[41]

**8.7.5.2. Affinity Chromatography.** In affinity chromatography, a specific inhibitor for an enzyme is immobilized on a support. When an enzyme mixture containing the desired enzyme is applied to the column, compounds that have no affinity for the inhibitor pass through. The desired enzyme remains bound to the column as a result of its attraction for the immobilized inhibitor ligand. The enzyme may then be eluted by a solution of free ligand or by changes in pH and/or ionic strength.[42] Scale-up of present laboratory procedures for affinity chromatography should considerably reduce the cost of many enzymes so that use of immobilized enzymes can be extended.

**8.7.5.3. Commercial Production.** Hydrolytic extracellular enzymes that do not require cofactors are being used commercially. A number of applications were reviewed by Skinner.[43] The largest commercial application of bound enzymes is the production of high-fructose corn syrups. Immobilized glucose isomerase converts D-glucose to D-fructose to provide a syrup that is equivalent in sweetness to sucrose and that is used extensively in canned fruits, fruit-flavoured beverages, carbonated drinks, and other products. The entire production starting from cornstarch could be accomplished with bound enzymes: the first—$\alpha$-amylase—catalyzes the breakdown of starch and can be used in place of acid hydrolysis; the second—$\alpha$-glucoamylase—catalyzes the breakdown of oligosaccharides and other carbohydrates

**Table 8.10.** Binding Mitochondria to Silylated Silica Beads[a]

| Alkyltrichlorosilane | % Hydrocarbon on bead | % Mitochondria, dry weight |
|---|---|---|
| Ethyl | 3.1 | 0.4 |
| Butyl | 2.7 | 0.4 |
| Octyl | 3.5 | 0.9 |
| Tetradecyl | 3.2 | 1.4 |
| Octadecyl | 3.0 | 1.3 |

[a] Porasil-B® (from Waters Associates): Will be silanized in pores, but mitochondria will be on the outer surface only.

to glucose; the third—glucose isomerase—converts glucose to the sweeter fructose.

**8.7.5.4. Whole Cells and Cell Organelles.** Complex enzyme reactions that require cofactors have been difficult to isolate in an active form in immobilized systems. One process might be to immobilize whole cells in which enzymes and cofactors are normally bound to the membrane and allowed to resemble more closely their in vivo conditions. Arkles et al.[44] reported on methods of binding whole cells and cell organelles to silylated surfaces. Because whole cells are sensitive to most reagents required for covalent bonding, immobilization was effected by noncovalent methods. Adsorption of cells and organelles through hydrophobic or ionic interaction is both mild and has little effect on substrate or metabolite diffusion.

Silica beads treated with octadecyltrichlorosilane were fairly effective in holding rat liver mitochondria (Table 8.10), rat liver microsomes, and spinach leaf chloroplasts. An ethylsilylated surface was preferred for binding erythrocytes. the binding process did not appear to have resulted in any functional changes in the organelles. Bound mitochondria utilized oxygen and ADP with a positive assay for ATP. Oxygen evolution was observed when bound chloroplasts were exposed to intense light, indicating that at least a portion of the photosynthesis apparatus remained intact.

## 8.8. Coated Metal Oxide Electrodes

Metal oxide semiconductor electrodes (e.g., $SnO_2$, $TiO_2$, etc.) may be coated with a monolayer of covalently bonded silane coupling agents. Electrochemical processes on such coated electrodes were first reported by Moses et al.[45] Electrochemical reactions in the presence of silane-treated oxides provided some insight into the completeness of the silylation reaction, as well as providing simple ways to modify charge-transfer reactions at an electrode. Diaz[46] reported on electrochemical reactions of pyrazolines bonded to an aminosilylated $SnO_2$ electrode as compared with pyrazolines in solution. Pyrazolines in solution undergo reversible one-electron reactions in contact with silylated (or unsilylated) $SnO_2$ electrodes, but when bound to a silylated surface, the pyrazolines undergo irreversible conversion to pyrazoles.

Oxidation of solution species in an electrochemical cell may be activated by ultraviolet light to generate an electric current. Various dye molecules in solutions in contact with a semiconductor electrode are capable of promoting an oxidation reaction sensitized by only visible-wavelength light. Armstrong[47] demonstrated that it was possible to accomplish such a catalytic cycle with the dye molecule attached to a silylated electrode. Xanthinetype dyes and metal phthalocyanines were bound to aminosilylated $SnO_2$ and $TiO_2$ electrode surfaces through amidization of carboxyl or sulfonyl chloride groups in the dyes.

Bound xanthine dyes yielded an equivalent photochemical response to that observed from an adsorbed dye, but with much better stability. Experiments conducted over several hours with a chopped visible light source showed no apparent degradation of the bound electrode response. Similar electrodes with electrochemically adsorbed dye showed irreversible loss of photocurrent over the same period of time. A cobalt phthalocyanine dye bound to a silylated $SnO_2$ electrode was sufficiently stable on the electrode to allow its redox state to be cycled over 1000 times without loss of electrochemical activity.

Only one monolayer of dye actively participates in the reaction, whether adsorbed from solution or covalently bound to the surface. A profitable area for research will be to vary the nature of the bond between dye and electrode in relation to electrochemical performance and stability. Lenhard and coworkers[48] have advocated considerable flexibility of the coupling molecule to allow reasonable molecular motion to accommodate electron transfer. More rigid bonds, especially multiple bonds, between dye and electrode would provide better stability. Orientation of the dye molecule also could be controlled by selecting reactive sites on the dye for bonding.

## 8.9. Liquid Crystals

In liquid crystal displays, clarity and permanence of image is enhanced if the display can be oriented parallel or perpendicular to the substrate. Oxide surfaces treated with octadecyl [3-(trimethoxysilyl)propyl] ammonium chloride, $C_{18}H_{37}^{\oplus}(CH_3)_2CH_2CH_2CH_2Si(OCH_3)_3Cl^-$, tend to orient liquid crystals perpendicular to the surface, while parallel orientation is obtained on surfaces treated with $N$-methylaminopropyltrimethoxysilane[49] $(CH_3NHCH_2CH_2CH_2Si(OCH_3)_3)$.

## References

1. D. E. Leyden and W. T. Collins, Eds. *Silylated Surfaces*, Gordon and Breach, New York (1980).
2. O. K. Johannson and J. J. Torok. *Proc. Inst. Radio Engrs.* **34**, 296 (1946).
3. E. Sacher, J. K. Sapieha, H. P. Schrieber, N. R. Wertheimer, and N. S. McIntyre. *Silanes, Surfaces, and Interfaces*, Vol. I, pp. 189-202, D. E. Leyden, Ed., Gordon and Breach, London (1986).
4. T. J. Nestrick, L. L. Lamparski, and R. H. Stehl. *Anal. Chem.* **51**, 2273 (1979).
5. F. W. Abel, F. H. Pollard, P. C. Uden, and G. Nickless, *J. Chromatogr.* **22**, 23 (1966).
6. P. Legally. U. S. Patent 3,108,920 (to Brockway Glass) (1963).
7. G. R. Bogart, D. E. Leyden, T. M. Wade, W. Schaefer, and P. W. Carr. *Chemically Modified Oxide Surface Symposium. J. Chromatogr.* **483**, 209 (1989).
8. C. Pidgeon, and U. V. Venkataram. *Analyt. Biochem.* **176**(1), 36 (1989).
9. L. E. Myers. *Soil and Water Conserv. J.* **22**(3), 95-97 (1967).
10. E. P. Plueddemann. *Proc. Water Harvesting Symp.* Phoenix, p. 76 (26 March 1974).
11. D. I. Hillel. U. S. Patent 4,027,428 (1978).
12. G. H. Snyder, H. J. Ozaki, and N. C. Hayslip. *Proc. Soil Sci. Soc. Am.* **38**(4), 678 (1974).
13. D. Scholl. *Proc. Soil Sci. Soc. Am.* **38**(4) (1974).
14. E. P. Plueddemann. U. S. Patent 4,557,854 (to Dow Corning) (1985).
15. C. H. Lochmuller and C. S. Amoss. *J. Chromatogr.* **108**, 85 (1975).
16. C. H. Lochmuller. In D. E. Leyden and W. T. Collins, Eds., *Silylated Surfaces*, pp. 231-267, Gordon and Breach, London (1980).
17. D. M. Hercules, L. E. Cox, S. Onisnick, G. D. Nichols, and J. C. Carber. *Anal. Chem.* **45**, 1973 (1973).
18. I. K. Guha and J. Janak. *J. Chromatogr.* **68**, 325 (1972).
19. F. C. Chow and E. Grushka. In *Silylated Surfaces*, D. E. Leyden and W. T. Collins, Eds., pp. 301-319, Gordon and Breach, London (1980).
20. F. K. Chow and E. Gruska. *Anal. Chem.* **50**, 1346 (1978).
21. D. E. Leyden. In *Silylated Surfaces*, D. E. Leyden and W. I. Collins, Eds., pp. 321-332, Gordon and Breach, London (1980).
22. D. E. Leyden and G. H. Luttrell. *Anal. Chem.* **47**, 161 (1975).
23. D. E. Leyden, G. H. Luttrell, W. K. Nonidez, and D. B. Werho. *Anal. Chem.* **48**, 67 (1976).
24. M. Debello, Master's Thesis, Univ. of Colorado (June 1981).
25. E. P. Plueddemann. In *Silanes, Surfaces, and Interfaces*, Vol I, pp. 1-24, D. E. Leyden, Ed. Gordon and Breach, London (1986).
26. E. P. Plueddemann. U. S. Patent 4,379,931 (to Dow Corning) (1983).
27. K. G. Allum, R. D. Hancock, I. V. Howell, S. McKenzie, R. C. Pithethy, and R. J. Robinson. *J. Organomet. Chem.* **87**, 203 (1975).
28. T. J. Pinnavaia, J. G-S. Lee, and M. Abedini. In *Silylated Surfaces*, D. E. Leyden and W. T. Collins, Eds. pp. 333-344, Gordon and Breach, London (1980).
29. E. P. Plueddemann and H. K. Chu. Unpublished data.
30. A. J. Isquith, E. A. Abbott, and P. A. Walters. *Appl. Microbiol.* **23**(6), 859 (1973).
31. W. C. White and R. L. Gettings. In *Silanes, Surfaces, and Interfaces*, Vol I, pp. 107-140, D. E. Leyden, Ed., Gordon and Breach, London (1986).

32. L. Blehm, J. R. Malek, W. C. White. U. S. Patent 4,631,273 (to Dow Corning) (1986).
33. W. Machleist. *Proc. Int. Conf. Solid Phase Methods in Protein Sequence Analysis*, p. 17 (1975).
34. M. Lynn. "Inorganic Support Intermediates: Covalent Coupling of Enzymes on Inorganic Supports." In *Immobilized Enzymes, Antigens, and Peptides*, H. H. Weetal, Ed., Marcel Dekker, New York (1975).
35. R. A. Messing. *Immobilized Enzymes for Industrial Reactors*, Academic, New York (1975).
36. (a) D. L. Eaton. In *Silylated Surfaces*, D. E. Leyden and W. T. Collins, Eds., pp. 201–227 Gordon and Breach, New York (1980). (b) D. L. Eaton. *Med. Res. Eng.* **12**(3), 17 (1976).
37. J. B. Peri. *J. Phys. Chem.* **701**, 2937 (1966).
38. R. D. Smith and P. E. Corbin. *J. Am. Ceram. Soc.* **32**(6), 195 (1949).
39. G. P. Royer and F. A. Liberatore. In *Silylated Surfaces*, D. E. Leyden and W. T. Collins, Eds., pp. 189–199, Gordon and Breach, New York (1980).
40. G. P. Royer. *CHEMTECH*, p. 695 (Nov. 1974).
41. R. S. Schrifreen, D. A. Hanna, L. D. Bowers, and P. W. Carr. *Anal. Chem.* **49**, 1929 (1977).
42. P. Creatrecasas and C. B. Anfinsen. *Ann. Rev. Biochem.* **40**, 259 (1971).
43. K. J. Skinner. "Enzymes Technology." In *Chem. Eng. News*, pp. 22–41 (18 Aug 1975).
44. B. C. Arkles, A. S. Miller, and W. S. Brinigar. In *Silylated Surfaces*, D. E. Leyden and W. T. Collins, Eds., pp. 363–375, Gordon and Breach, New York (1980).
45. P. R. Moses, L. Wier, and R. W. Murray. *Anal. Chem.* **47**, 1882 (1975).
46. (a) A. Diaz. *J. Am. Chem. Soc.* **99**, 5383 (1977). (b) A. Diaz. In *Silylated Surfaces*, D. E. Leyden and W. T. Collins, Eds., pp. 137–157, Gordon and Breach, London (1980).
47. N. R. Armstrong. In *Silylated Surfaces*, D. E. Leyden and W. T. Colins, Eds., pp. 159–171, Gordon and Breach, New York (1980).
48. J. R. Lenhard and R. W. Murray. *U. Electroanal. Chem.* **77**, 393 (1977).
49. F. J. Kahn, G. N. Taylor, and H. Schonhorn. *Proc. Inst. Electr. Electron. Eng.* **61**, 823 (1973).

# Index